Data Strategy Reconsidered

Seven Principles That Challenge Convention and Deliver Results

Kevin Lewis

ISBN: 979-8-9938589-1-3

This book is dedicated to my son, Jack,
whose curiosity, passion, and courage
delight and inspire me every day.

CONTENTS

ACKNOWLEDGMENTS

This book exists because of the collective wisdom of hundreds of colleagues and clients who shared in the journey to discover what actually works in data strategy through hard-won experience.

I'm deeply grateful to my colleagues at AWS Professional Services, where I led the development of the Modern Data Strategy methodology that informs much of this book. Your bias for action and insistence on the highest standards made these principles possible.

To my former colleagues at Teradata who taught me to recognize universal patterns across industries and showed me that effective enterprise-level programs and capabilities share common success factors, regardless of the organization.

My foundation in data strategy was built at Publix Super Markets, where I initiated and led our enterprise data analytics program. Thank you for giving me the opportunity to learn by doing and for showing me how organizations really work from the ground up from my earliest days as a bagger in high school and throughout my career at Publix.

Although much of this book is based on my experience with these companies, the opinions expressed are my own and not necessarily those of any former or current employer.

Most important, to my clients over the years: thank you for trusting me with the unvarnished truth about your challenges along with your successes. While confidentiality prevents me from naming you, your fingerprints are all over these principles. You wanted things fixed, not just analyzed, and that urgency drove the development of approaches that actually work.

To everyone who has demanded practical solutions over impressive theories—this book is for you.

INTRODUCTION: A BETTER PATH TO DATA STRATEGY SUCCESS

If you're reading this book, you're probably working on a data strategy—either building one from scratch or improving one that isn't delivering the results you'd hoped for. Either way, you're in good company. Over the past 25+ years, I've been on this same journey many times, first as a practitioner building my own data and analytics program and later as a consultant helping hundreds of organizations transform their data capabilities.

Here's what I've learned: While data strategy is challenging, it's absolutely achievable when you approach it with the right principles. The problem is that most of the conventional wisdom about data strategy sounds perfectly reasonable but doesn't quite work in practice. Organizations follow this advice faithfully, invest significant time and resources, yet still struggle to deliver the transformative results they're seeking.

The Hidden Pattern

When I first built my own data and analytics program, I followed the advice in the most popular books on data strategy available at the time. Some of these books provided excellent theoretical frameworks, while others offered detailed and useful technical advice. But when it came to actual enterprise-wide implementation—the messy reality of working inside large organizations—their guidance often fell short. It became clear that much of the strategic advice available wasn't based on long-term, hands-on experience with the complexities of big companies or government agencies.

So, I learned through trial and error. I experimented with different approaches, adjusted when things weren't working, and gradually discovered principles that delivered results. Some of these discoveries surprised me because they contradicted much of the conventional wisdom I'd been reading. But they worked—consistently and reliably.

It wasn't until I became a consultant and started working with many different organizations that I truly appreciated what I'd learned earlier in my career. As I helped companies across every major industry with their data strategies, I was struck by something remarkable: nearly all of them were struggling with the exact same challenges. Despite their different industries, cultures, and technologies, they were making the same fundamental mistakes and hitting the same roadblocks.

A financial services company would be struggling with data silos and lack of business engagement. A manufacturer would show up the same issues. A healthcare organization, a retailer, an agricultural firm—different businesses, identical problems. And I found myself giving remarkably similar advice to each of them, based on what I'd learned through my own trial and error.

These weren't random, unrelated issues. There were clear patterns. Organizations were following what seemed to be consensus guidance, and it was consistently leading them astray. They were trying to build business cases based on the proposed direct value of their data strategy. They were building "foundations" for the enterprise, attempting to anticipate all possible requirements. They were swinging between extremes of centralization and decentralization, often with both extremes existing at the same time in the same organization. The same mistakes, over and over.

The frustrating thing is that these challenges and the root causes haven't been communicated widely enough. The popular books and articles are still promoting the same theoretical approaches that sound good but don't work in practice.

The Seven Principles That Change Everything

Through years of hands-on experience and observation, I've identified seven principles that distinguish successful data strategies from those that struggle. These aren't theoretical concepts. They're practical, proven approaches that work in the reality of large organizations. They're applicable to all large enterprises: public and private, new and old, technically advanced and neophyte alike.

Principle #1: Target Business Initiatives, Not Direct Business Value

This first principle might surprise you. Everyone says to start with business value to build a business case for your data strategy and drive the program. It sounds logical. But here's what successful organizations have discovered: Instead of trying to create value directly through your data strategy, align it to specific, funded business initiatives that are already underway or planned. Show value by *supporting* the value to be produced by these business initiatives.

This subtle shift changes everything. Rather than competing for attention and resources, you become an enabler of the organization's most important priorities. Rather than hoping stakeholders will see your value, you demonstrate it through *their* success. It's a powerful reorientation that immediately increases your relevance and impact.

Principle #2: Build for Change, Not for All Possible Requirements

When you focus on specific business initiatives, a natural concern arises: "Won't we have to rebuild everything when new requirements come along?" Successful organizations have found a better way—building with extensibility in mind from the start.

This doesn't mean trying to anticipate every possible future requirement, which is impossible and leads to over-engineering. It means following specific guidelines that ensure what you build today can be extended tomorrow. Structure data at its most granular level. Source from systems of record. Build adjustable pipelines. These practices ensure that serving today's known needs creates a foundation for tomorrow's unknown needs.

Principle #3: Balance Responsibility, Don't Polarize

Organizations often swing between extremes of centralization and decentralization in their data capabilities. The successful ones have learned that neither extreme is optimal. Complete centralization can create bottlenecks, while complete decentralization can lead to fragmentation.

The key is thoughtfully balancing responsibilities between central teams, data producers, and data consumers based on what your business initiatives require. Some initiatives benefit from enterprise-wide coordination, while others can be handled within domains. The organizations that get this right match their structure to their business needs, not to abstract organizational theories.

Principle #4: Build Coherent Architecture, Not Monolithic Data Stores or Disconnected Data Products

To enable data integration and sharing throughout the organization while also distributing responsibilities effectively, you need avoid building a single enterprise monolith on one extreme while also going beyond a large inventory of disconnected data products on the other extreme. You need data that actually connects when business initiatives require true linkage across domains while *also* allowing distributed teams to contribute to that integration. Having excellent individual data products means little if they can't work together when needed by business initiatives.

This means establishing semantic linkages across domains and ensuring that frequently-joined, high-volume, widely-used data is physically co-located for performance. It's about building an architecture where the Raw layer preserves domain reality, the Harmonized layer creates both semantic and physical integration where required, and the Curated layer optimizes for specific uses. The goal isn't to constrain domains but to ensure coherent integration is possible when business initiatives demand it.

Principle #5: Focus on the Future, Not on Rewriting Legacy

Every organization has technical debt including redundant systems, duplicate data, and inconsistent structures built up over years or decades. The temptation is to clean it all up before moving forward. But the organizations that succeed take a different approach.

Instead of diving into the mess, they parse the work into three distinct tracks: business initiative-driven new development, technology-only migration, and surgical fixes. This allows them to build the future while managing the present, delivering value immediately while gradually retiring legacy systems through natural evolution rather than massive rewrite projects.

Principle #6: Plan the Program, Not Just Individual Solutions

Even with good alignment and careful attention to extensibility, you need coordination to achieve coherent results. The most successful organizations treat data strategy as a program, not just a collection of projects.

This means having a disciplined framework with continuous and iterative phases. This includes identifying which initiatives to support, assessing the program to understand what's needed, planning a coordinated roadmap, and implementing to deliver value while simultaneously building your foundation. This program approach ensures every project contributes to a larger vision while delivering immediate value.

Principle #7: Embed Data Strategy, Don't Isolate It

This final principle determines whether your data strategy becomes a permanent capability or remains a constant struggle. Successful organizations weave data considerations into their existing enterprise operating model including their strategic planning processes, funding mechanisms, architecture governance, and development methodologies.

This isn't about creating bureaucracy. It's about making coherent data planning and implementation a natural part of how your organization operates. When data strategy is truly embedded, business initiatives automatically consider data requirements, project teams naturally leverage shared resources, and good data practices persist through organizational changes.

Why These Principles Work

What makes these principles so effective? Three things:

First, they're based on real-world success patterns, refined through experience across hundreds of organizations. When I suggest that you align to business initiatives rather than direct value, it's because I've seen this approach consistently accelerate success. When I emphasize building for extensibility, it's because I've experienced the results of this approach myself, and I've watched organizations that do this adapt smoothly to new requirements while others struggle with constant rework.

Second, these principles work together as a system. Each reinforces and depends on the others. Aligning to business initiatives becomes more powerful when you build for extensibility. Balancing centralization and decentralization works better within a program framework. The program approach succeeds when embedded in the operating model. Together, they create a sustainable approach to data strategy.

Third, this approach works with organizations as they are, not as we wish they were. It doesn't require massive transformation or restructuring. It works within existing constraints and politics. It builds on what you already have.

A Practical Approach

This book focuses on practical application, not abstract theory. You won't find maturity models or capability frameworks here. While those tools have their place, I've seen too many organizations focus on achieving high maturity scores without delivering meaningful business results.

You also won't find specific technology recommendations. I won't tell you which database to choose, or which data management tools work best. Technology choices matter, but they're not what primarily determines success. Organizations with identical technology stacks can have vastly different outcomes based on how they approach their strategy.

And while these principles absolutely apply to establishing a data foundation for AI, I won't get into detail on this connection. But be assured that AI applications and associated enterprise-wide transformation require this type of thinking for data strategy more than ever.

What you will find, instead, are specific, timeless, actionable principles you can start applying immediately, regardless of your current situation or technology choices.

How to Use This Book

Each chapter that follows explores one principle in depth, showing you not just what to do, but why it matters and how to do it. I'll share real examples (with some details changed to protect confidentiality) that illustrate both the challenges and the solutions. Most importantly, I'll give you specific, practical guidance you can apply immediately.

I'll finish up by answering some of the most frequent questions I get from clients.

Looking Forward

I'm excited to share these principles with you because I know the difference they can make. I've seen organizations go from struggling with basic data integration to enabling a wide variety of sophisticated applications. I've watched data teams transform from isolated cost centers to strategic partners. I've witnessed the satisfaction of business leaders who finally have the data capabilities they need to compete effectively.

The path to data strategy success isn't always easy, but it is achievable. You don't have to figure everything out through trial and error. You can learn from the experiences of hundreds of organizations that have walked this path before you.

The data-driven future that everyone talks about isn't just possible, it's within reach when you approach it the right way. Let's explore how to make it reality for your organization.

PRINCIPLE #1: TARGET BUSINESS INITIATIVES, NOT DIRECT BUSINESS VALUE

- ✓ *Do:* Align data strategy to funded business initiatives
- × *Don't:* Align data strategy directly to business value

Let me start with the most important principle of all—and the one that will save you literally years of struggle.

This principle goes against 90+% of the advice you'll find on data strategy, and yet not following this principle is by far the most common reason for a slow and painful failure.

I'm not exaggerating even a little. After more than 25 years and hundreds of clients, I've seen the same pattern play out repeatedly: organizations struggling with their data strategy because they're following advice that sounds right but simply doesn't work.

The Conventional Wisdom Trap

If you were to do a Google search, ask your favorite AI chatbot, or consult most data strategy experts, seeking advice on how to develop a data strategy, they'd all almost certainly advise you to do something like this:

1. Start by making a business case for your data strategy; that is, identify the business value you intend to produce (of course, right?)

2. Assess your current state data and associated capabilities and envision the future state.
3. Implement incrementally (don't take the "big bang" approach).
4. Measure value at each iteration.

Then, at the end of each iteration, business initiatives can leverage the high-quality data you've made available.

Sounds reasonable, right?

Here's what I can tell you after witnessing the results of this approach over and over again: *It does not work.* I don't care who recommends it, how credible they seem, or how many different people say it. It fails. Consistently.

The Fatal Flaw of the Value-First Approach

Here's what happens when you try to create business value directly through your data strategy: While you're building your theoretical foundation, every major business initiative in your organization still needs data. What do they do? They build their own data solutions—independently, redundantly, and inconsistently.

You end up with the worst of both worlds: a data team building platforms that are severely underutilized and untrusted, and business teams creating a fragmented data landscape because they had no other choice.

The value proposition in this anti-pattern typically takes the form of "making it easier for end users to access high-quality data," often accompanied by calculated return on investment figures. While this outcome would of course be beneficial, it has little to no connection to the enterprise's core business activities and consequently fails to generate enthusiasm among the executives expected to fund such initiatives. If they're lucky, they don't get the funding they want, forcing them to re-think their approach. If they're unlucky, they get the funding and end up figuring it out the hard way, if they figure it out at all.

The more sophisticated version of this anti-pattern does propose business value that directly addresses the actual business. In an insurance company, for instance, you might argue that your data strategy will enable more effective claims adjudication. Yet this approach remains fundamentally flawed. Why? Because there are—or should be—initiatives already sponsored by the executive responsible for claims operations and funded based on their own merit, not through funding associated with data strategy.

Let me illustrate this with an example. Years ago, while working with an automobile manufacturer, our team interviewed various stakeholders across the organization. During our meeting with a marketing executive, he enthusiastically presented what he considered a highly valuable opportunity. His idea involved targeting customers who had contacted the company for non-sales reasons, such as warranty issues. He produced a spreadsheet demonstrating impressive ROI projections. Seems like a fantastic opportunity, doesn't it? Yet had we proceeded to support this idea, it would have been a complete waste of time. The executive inadvertently revealed why when he said, "We've been proposing this idea for the last three years and can't get it approved. I'm glad you have funding available for the data strategy, so we can finally get this work done."

If this initiative couldn't secure approval alongside other marketing proposals through normal channels, why would it suddenly become a sound target for the data strategy? And what about the initiatives that this executive *did* have approved? How would they acquire the data they needed for success?

The Business Initiative-First Approach

The alternative I'm proposing might seem like a subtle shift, but the results are anything but subtle, they're dramatic. Here's the move that changes everything:

Put business initiatives (not direct value) at the front of the line.

That's it. Instead of starting with "business value" to be created by the data strategy itself, start with specific, funded business *initiatives*. This single change is the difference between organizations struggling endlessly with their data strategy and those producing real results that matter.

Here's an approach that works much better than the value-first approach:

1. Identify the major business initiatives sponsored outside of the data organization that are already underway or planned and funded (or likely to be funded) such as customer experience transformation, supply chain optimization, and so on. (Of course, you should educate initiative sponsors about how data can be used in innovative ways for their business areas, so that *they* can propose these ideas within *their* initiatives, not yours.)

2. Determine the data these initiatives need to succeed. Here, you'll discover that many initiatives need similar data in slightly different forms, allowing you to plan for reuse.
3. Implement data management capabilities specifically to address quality and other issues with the data required by these initiatives and specific applications. As you support multiple initiatives, you'll naturally build reusable capabilities such as data quality management, master data management, data integration, data lineage, security, and so on.
4. Assess your *contribution* to the success (and hence value) of the targeted business initiatives. This is how you show *real* value, even if measuring that value may be imprecise.

A Structured Framework[1]

To make this principle more concrete, consider these seven workstreams as a checklist for your data strategy:

- *Business Initiatives:* The funded, long-running programs containing multiple projects. Which business initiatives (sponsored outside of the data organization) is your strategy committed to supporting with underlying data needs?
- *Applications / Use Cases:* Individual applications within initiatives (recommendation engines, forecasting algorithms, agentic workflows, analytics dashboards, and so on). Which specific data-intensive application projects within the targeted business initiatives will your data strategy support in the near-term and over time?
- *Data:* The specific data required to support targeted applications. What data domains (e.g., customer, product, sales) including structured and unstructured data, and internal and external data, will be needed to enable targeted initiatives and applications?
- *Data Management:* Capabilities ensuring data readiness (data quality management, master data management, data lineage, data catalog, etc.) What specific data issues associated with the above data domains will hinder success of the targeted business initiatives and associated application projects if not addressed? How will data consumers find and understand the

[1] I led the development of this framework as part of the Modern Data Strategy methodology from AWS Professional Services (ProServe).

data they need? Which capabilities will be needed to address the issues?

- *Architecture:* Technical components for processing, storing, and sharing data. What technologies will be required to manage the needed data and enable the targeted applications? How will all the components fit together?
- *Security and Responsibility:* Protection from unauthorized access and use. What new security, regulatory, policy, or ethical issues will you encounter in the targeted business initiatives and applications? How will they be addressed?
- *Operating Model:* People and processes for managing and leveraging data. What non-technical organizational structures, roles, and processes will be needed to manage, share, and use the data, such as data governance, data stewardship, analytical, and data science capabilities? How will you continuously update your plans based on new and evolving business initiatives?

Business Initiatives	Which business initiatives is the data strategy supporting?
Applications / Use Cases	Which applications will be deployed within the target business initiatives?
Data	What data will be required for the targeted applications?
Data Management	What data issues will need to be resolved?
Data and AI Architecture	Which elements of the architecture will need to be built or enhanced?
Security and Responsibility	What security-related risks will need to be managed?
Operating Model	How will business and IT roles and processes be optimized?

Data Strategy Workstreams – These seven workstreams can be used as a checklist to ensure you have all the right elements within your data strategy.

Your data strategy should be able to answer the above questions confidently. If you can't, then something important is missing.

Two Strategic Mistakes to Avoid

Upon first glance, these workstreams might seem obvious, but slight variations in the framework reveal two devastating patterns I see repeatedly:

Strategic Mistake #1: Data Proliferation

Strategic Mistake #1 is when organizations skip the data workstream entirely, leaving each application team to source their own data. The result is widespread proliferation including redundant pipelines, overlapping structures, and inconsistent data everywhere. Each team builds complete solutions for their specific needs without considering the broader landscape.

Business Initiatives	Which business initiatives is the data strategy supporting?
Applications / Use Cases	Which applications will be deployed within the target business initiatives?
Data	What data will be required for the targeted applications?
Data Management	What data issues will need to be resolved?
Data and AI Architecture	Which elements of the architecture will need to be built or enhanced?
Security and Responsibility	What security-related risks will need to be managed?
Operating Model	How will business and IT roles and processes be optimized?

Strategic Mistake #1 – This mistake happens when each application project directly collects data from required sources, leading to data proliferation.

Strategic Mistake #2: The Disconnected Foundation

Here's what's tragic: When organizations realize they're making Strategic Mistake #1, they almost always (and I do mean *almost always*) overcorrect into Strategic Mistake #2. That is, they disconnect from business initiatives entirely, reasoning that close alignment to initiatives is what caused the proliferation problem in the first place. So, they set out to build a "data foundation" for everyone.

Business Initiatives	?
Applications / Use Cases	?
Data	What data will be required for the targeted applications?
Data Management	What data issues will need to be resolved?
Data and AI Architecture	Which elements of the architecture will need to be built or enhanced?
Security and Responsibility	What security-related risks will need to be managed?
Operating Model	How will business and IT roles and processes be optimized?

Strategic Mistake #2 – This mistake is an extremely common over-correction of Strategic Mistake #1. In an effort to build a foundation for all possible uses, the data team disconnects from specific business initiatives and applications.

But without connection to specific initiatives, how do you prioritize your work? For example, which data should be deployed first, second, and third? Which quality issues matter most? You can survey users, but they're just sharing their individual opinions on priorities or what affects them most directly.

With this mistake, the work drags on with no clear value delivery, enthusiasm wanes, and eventually the program thrashes around for a new approach, sometimes switching back to Strategic Mistake #1 again.

Variations on the Mistakes

These mistakes can occur in multiple forms, associated with various workstreams. For example, I've lost count of how many organizations develop data governance operating models in near complete isolation, separate from any data strategy, disconnected from business initiatives, with their own vague value propositions. These programs launch with fanfare but ultimately fizzle out, slowly and quietly. Why? Because despite all the governance frameworks and policies, they're not actually helping anyone succeed at anything that matters. This happens even as individual application projects address data quality issues on their own—issues that data governance is supposed to help with—in parallel to the struggling data governance program.

The lesson should be clear: Don't align your data strategy and associated capabilities to business value directly. Align it to specific, named, funded business initiatives that are sponsored outside of the data organization—without burying data and associated capability delivery completely within those initiatives. The value will follow because you're supporting the most carefully scrutinized value propositions of the enterprise. These initiatives would proceed and would acquire the data they need, with or without your help. And if you do help, you can do so in a way that simultaneously builds the kind of foundation you were aiming for in Strategic Mistake #2 while adding value to targeted application projects like Strategic Mistake #1. In other words, you're finding the right middle ground between these two common mistakes.

Thinking Big, Starting Small

By aligning your data strategy actions to specific business initiatives, each implementation will accomplish two things: First, it will deliver business value in small scope projects, where every action is focused on ensuring success of the targeted application projects. Second, it will contribute to a coherent data landscape, like contributing a single puzzle piece to the larger enterprise puzzle.

But wait, if we're only deploying data and associated capabilities in small-scope efforts, focused on targeted applications, what happens when new applications come along and need the same data but in a slightly different form? If we want reusability, won't we have to re-design our data pipelines and data structures to accommodate new data? If we don't, won't we end up proliferating data again?

Principle #2 addresses these questions.

PRINCIPLE #2: BUILD FOR CHANGE, NOT FOR ALL POSSIBLE REQUIREMENTS

✓ *Do:* Deploy data with extensibility
× *Don't:* Deploy data as a "foundation" for all possible application needs

The first questions I usually get when advising clients to focus on specific business initiatives and applications, as outlined in Principle #1, are: "If we implement our data that way, won't we paint ourselves into a corner? What happens when a new application comes along that needs the same data, but with different requirements? Won't we end up either creating redundant data pipelines or completely rewriting what we've already built?"

The answer is no, but you're absolutely right to worry about it. Preventing this situation requires some forethought and discipline.

The solution is to follow specific guidelines that ensure extensibility of your data architecture. It's far better to design for seamlessly adding to your platform than to attempt to understand all possible future requirements at the outset. Doing incremental new data work for new applications is perfectly fine, but doing extensive rework is not.

Let's examine the specific guidelines that enable true extensibility.

Structure data at the lowest level of detail, not just summaries.

If you're deploying data for a specific application, that application might only require aggregated data. But here's the trap: optimizing solely for the immediate need will force you into rework or duplication later.

For example, consider a global video streaming platform launching a personalized recommendation engine. The initial application might only need daily viewing metrics by user and content category. You'd be tempted to extract viewing data, aggregate it to daily summaries, and deliver exactly that. But what happens when the fraud detection team needs the same viewing data but requires individual session details to identify suspicious patterns? Or when the content team wants minute-by-minute engagement data to optimize show pacing?

Suddenly you're building separate pipelines for each use case, which is exactly the proliferation problem we're trying to avoid.

Instead, structure your data at the most granular level from the start. Even though the recommendation engine needs daily aggregates, your data workstream should capture individual viewing events with full session details, user interactions, and temporal precision. The recommendation engine gets its daily summaries through transformation, but now fraud detection can access session-level details and content optimization can drill down to engagement patterns from the same foundational dataset.

Obtain data from original sources, not secondary sources.

The original source, or "source of record," is where data first enters your organization. Secondary sources are everywhere that data travels afterward. While it's often easier to grab data from a secondary source, this creates hidden dependencies that will sabotage future applications.

Take a pharmaceutical company tracking drug trials. Patient data might originate in clinical trial management systems, then flow to various application-specific analytics platforms such as regulatory reporting databases, safety monitoring systems, and other solutions. When building a machine learning model to predict trial outcomes, for example, it would be tempting to pull data from one of the application-specific analytics platforms into the enterprise data platform. It's already cleaned and may be structured in a manner that suits your needs or can be easily transformed to do so.

But here's what goes wrong: The next application might be a real-time safety monitoring system that needs immediate access to adverse event reports. The analytics platform you chose to source data from might have a 24-hour delay and exclude certain data points deemed non-essential for its targeted reporting. Now you need a completely separate pipeline from the original clinical systems.

Always go to the source of record when acquiring data for a shared data platform. Yes, it requires more initial coordination with source system owners, but it prevents the cascading complexity that destroys reusability. And when multiple systems serve as sources of record for the same business entity, such as patient data entered in both trial management and hospital systems, for example, involve data stewards to establish authoritative sources for specific attributes.

Build right-time and adjustable data pipelines, not rigid ingestion processes.

Different applications need different levels of data timeliness, but building everything for real-time processing "just in case it will be needed" is prohibitively expensive. The key is building pipelines that can evolve their timeliness characteristics without complete reconstruction.

Consider an e-commerce company starting with a daily sales analysis dashboard. The temptation is to build a simple nightly batch process—extract sales data, transform it, and load it into the enterprise data platform. But what happens when the marketing team wants to launch real-time personalization that needs purchase data within minutes?

Instead of building everything as batches, implement an event-driven architecture from the start. When a purchase occurs, publish that event with rich context including customer details, product information, pricing, timing, and session data. Your initial daily dashboard can consume these events in scheduled batches, but the infrastructure is already in place for real-time consumption in the future with less extensive changes required.

When the personalization requirement emerges, you don't rebuild the entire pipeline. You modify the pipeline to increase the timeliness without the need to create an additional connection to the source system. The beauty is that you can adjust timeliness incrementally. Maybe start with hourly batches for inventory management, then move to 15-minute intervals for demand signaling, and finally real-time for personalization, all using data sourced from the same underlying event streams.

This approach works equally well with change data capture (CDC) technologies, where you can adjust the frequency and sophistication of change detection as requirements evolve.

Acquire as much data as possible from sources, but only refine what's needed by targeted use cases.

When teams connect to enterprise data sources, they face a critical decision about scope: should they extract only what's immediately needed, or capture everything available? This guideline encourages comprehensive acquisition at the point of extraction. By landing all available data from a source system into the Raw layer—even fields and tables not required by current use cases—organizations create a valuable reservoir of unrefined data that can be tapped for future needs without returning to the source systems. (We'll explore the Raw, Harmonized, and Curated layers in detail when we discuss Principle #4.)

This approach delivers a crucial operational benefit that becomes apparent over time. When new applications or analytic use cases emerge requiring additional data from an already-connected source, teams can bypass the often complex and time-consuming process of re-establishing source connections, negotiating extraction windows, and managing incremental loads. Instead, they simply identify the needed data already residing in the Raw layer and focus their efforts on refining it through the Harmonized and Curated layers. This eliminates redundant integration work and accelerates time-to-value for new initiatives.

Consider a scenario where a finance team initially connects to the general ledger system to extract data for monthly reporting. Even if they only need specific account balances and journal entry summaries, following this guideline, they would extract all available general ledger data including every account, transaction detail, audit trail, and supporting table. Months later, when the audit team requires detailed transaction-level data with approval workflows, or when finance needs to analyze cost center allocations, that data already exists in the Raw layer, waiting to be refined rather than requiring new data extraction from the source.

The key insight is recognizing that the effort required to extract and land data in the Raw layer is relatively modest compared to the intensive work of refinement, which includes applying business rules, ensuring data quality, standardizing formats, and maintaining ongoing governance. By front-loading the acquisition and being selective about refinement, organizations optimize their resource allocation. They avoid the repeated cost of source system connections while investing refinement efforts only where demonstrable business need (and funded scope) exists. This creates a data platform that can respond quickly to new requirements while maintaining manageable operational overhead.

Correct data quality issues at the source of record, not the destination.

When you start using operational data for analytics and AI applications, data quality issues that were invisible or unimportant in transactional systems suddenly become glaring problems. The temptation is to fix these issues in your enterprise data environment because it seems faster and doesn't require coordination with source system owners. This is a dangerous shortcut that creates long-term chaos.

Imagine a logistics company building predictive models for delivery optimization. They discover that address data in their order management system is inconsistent. Maybe some entries use abbreviations, others don't, and postal codes are sometimes missing. The easy fix is to clean this data in the data ingestion pipeline using address validation services.

But now you have two versions of "truth"—the original messy data in operational systems and the cleaned data in the enterprise data platform. When the customer service team tries to reconcile delivery issues, they're looking at different address information than the predictive models. When the billing system generates invoices, it uses the original messy data, creating customer confusion.

Worse, the enterprise data ends up contradicting the business context reflected in operational systems. Maybe those abbreviated addresses represent specific delivery instructions that drivers understand, or postal code variations indicate different service zones. People come up with all kinds of creative but unintuitive ways to use data fields.

Instead, implement data quality monitoring in your enterprise data environment, but feed quality issues back to source system owners for correction. This creates a virtuous cycle: operational systems get cleaner over time, analytic applications get higher quality data, and everyone works from the same version of truth.

Build data models from conceptual to logical to physical, not just physical.

Jumping straight to physical data structures based on immediate application needs and expecting those data structures to be reusable and integrated across applications is like building an apartment complex by just slapping down some concrete and stacking up bricks.

You might make (apparent) progress faster, but you'll create a mess that's very unlikely to be the multi-tenant structure you intended.

Consider a financial services company building a customer analytics application. The quick approach is to source data from the CRM database structure, build tables in the analytics environment, and start building dashboards. This works initially but creates problems as the ecosystem grows.

When they add transaction analysis, they discover customer data is also in the core banking system with different structures and relationships. When they add investment portfolio analysis, customer data appears again in the wealth management platform with yet another schema. Without a coherent model, they end up with multiple, conflicting customer representations that can't be easily integrated.

Instead, start with a cross-domain conceptual model—a high-level view of how your key business entities relate to each other. For the financial services company, this might define customers, accounts, products, transactions, and their relationships. This shouldn't take months—a few weeks of collaborative modeling with business stakeholders should do it.

Then build logical models for specific domains as applications need them, detailing the attributes and relationships required for targeted use cases. Finally, implement physical structures optimized for performance while maintaining the ability to join data across domains according to your conceptual model.

This approach creates a coherent foundation that applications can build upon, rather than a collection of application-specific silos that resist integration.

Build data models based on stable business entities, not source structures.

Modeling data based on source system structures is like organizing your library based on which bookstore you bought each book from. It might reflect how the data arrived, but it doesn't reflect how people actually think about or use that information.

Take a healthcare system implementing population health analytics. Their electronic health record (EHR) system stores patient data in hundreds of tables reflecting the software's internal structure such as demographics in one area, visits in another, and diagnoses scattered across multiple tables based on coding systems. If they simply replicate this structure in their analytics environment, every application must understand the EHR's internal complexity.

When they add data from laboratory systems, pharmacy systems, and insurance claims, each brings its own structural quirks. Applications end up needing complex logic to navigate these various structures, and any change in source systems cascades through all dependent applications.

Instead, model around stable business entities that reflect how healthcare professionals think about their work: patients, providers, encounters, treatments, outcomes, and so on. These concepts remain stable even as underlying systems change. When the organization switches EHR vendors, only the transformation logic needs to be updated. Applications that leverage the enterprise data platform continue working with the same logical patient, encounter, and treatment entities.

This business-centric modeling also makes data more accessible to non-technical users who understand patients and treatments but shouldn't need to understand the internal structure of software systems.

Assign data stewards to business-based data domains, not systems.

Having system experts is valuable, but data stewardship organized around systems creates silos that work against enterprise coherence. You need stewards who think across systems to ensure consistent, enterprise-wide data integrity.

Consider a retail organization where customer data exists in e-commerce platforms, point-of-sale systems, loyalty programs, and customer service applications. If stewardship is organized by system, each steward optimizes for their specific system's data needs. The e-commerce steward focuses on web conversion metrics, the point-of-sale steward on transaction processing, and the loyalty steward on program engagement.

But what happens when you need a unified view of customer behavior across channels? Without enterprise-level stewardship, you get conflicting definitions. Different systems define "active customer" differently, attribute purchases differently, and handle returns through incompatible processes.

Instead, assign stewards to business domains that cut across systems. A customer data steward takes responsibility for ensuring consistent customer definitions, standardized attribution rules, and coherent customer lifecycle management across all systems. A product data steward ensures that the same product has the same set of core attributes whether it's sold online, in-store, or through mobile channels, even if the product has additional attributes unique to each channel.

You might organize stewardship around core business entities (customers, products, transactions, locations) or business segments (retail operations, digital channels, supply chain) or, more likely, a combination of both so that core data stewards promote cross-domain consistency through common data elements while domain stewards align to those core standards and linking them to unique data within their business domain. The key is that stewards have enterprise-wide accountability for their domain's data integrity, motivating them to contribute to coherence rather than focusing only on individual systems.

From Chaos to Clarity: A Retail Data Transformation.

When I was responsible for enterprise data and analytics at a large retailer, we faced a challenge that I now recognize as quite common. Each time we needed to build an analytical solution, we approached it as an isolated project with its own dedicated data, pipelines, and infrastructure. For example, when we built a store profitability analysis system that needed sales data, we constructed a custom pipeline from the point-of-sale systems. When the category management team required similar information, we built an entirely separate pipeline. Ad-hoc business requests meant yet more data extracts from the operational systems.

This pattern repeated across every business function including product management analytics, distribution optimization, order processing insights, and many others. We pulled data redundantly from the same core systems repeatedly, creating a web of duplicate pipelines that were expensive to maintain and difficult to govern.

If we were to build truly reusable data, I realized we needed to shift our perspective. Instead of asking "What specific data does this application need?" I began asking "What is the most granular, useful form of this data that could serve multiple purposes?"

For sales data, the answer was individual transaction records. Rather than immediately summarizing data for a new inventory replenishment system that needed sales by product, store, and day, we preserved each transaction in the enterprise data platform. This single decision transformed our entire approach.

When the inventory replenishment system required its daily summaries, we aggregated the transaction data accordingly. Then, when a labor planning application arrived months later needing sales data by department, store, and 15-minute intervals, the same underlying transactions supported this new view. Price optimization, shelf-space planning, and loss prevention systems all drew from the same well-organized, granular dataset.

The impact extended beyond our planned applications. Ad-hoc business requests which previously requiring custom extracts from production systems could now be satisfied using clean, organized data that was already available and understood.

By designing data assets to be inherently reusable and extensible rather than application-specific, we dramatically reduced redundancy while increasing agility. The key insight I gained was understanding that today's specific requirement can be positioned as a building block for tomorrow's applications, even if we can't predict what those applications will be.

Guiding Projects

With these guidelines in place, you need to ensure each data delivery project follows them consistently. Build this guidance directly into your project charters, epic definitions, or whatever documents you use to set objectives and constraints for projects.

Beyond referencing the principles, ensure that data delivery projects have a clear dual objective: first, deliver data as needed by target applications—no more, no less—and second, contribute to a shared, coherent data environment that enables future reuse.

This is why separating the data workstream from application workstreams, as discussed in Principle #1, is so critical. The data teams can maintain these dual objectives while application teams focus solely on building their specific solutions. This division of responsibility ensures that data teams not only follow the guidelines outlined in this principle but also find other ways to ensure data connects together coherently.

But what happens when multiple teams are contributing data independently? How do you enable that independence while ensuring standards are followed closely enough to build a genuinely coherent data landscape?

Let's discuss that in Principle #3.

PRINCIPLE #3: BALANCE RESPONSIBILITY, DON'T POLARIZE

- ✓ *Do:* Carefully balance centralization and decentralization of responsibilities
- × *Don't:* Swing responsibilities to extremes in either direction

One of the most persistent challenges in data strategy is finding the right balance between centralization and decentralization of responsibilities. Over the years, organizations have swung like a pendulum between these two extremes, overcorrecting in both directions and missing the nuanced middle ground that applies the best of both.

The Historical Pattern

In the early days of what at the time was simply referred to as "data warehousing," the prevailing wisdom was to centralize just about everything. Organizations had been drowning in fragmented data scattered across numerous departmental systems and splintered "data marts." The solution seemed obvious: establish a central team to bring all that data together into a single, centralized repository where it could be standardized, integrated, and reused across the enterprise.

This centralized approach worked well initially. Data became more consistent, cross-functional analytics became possible, and organizations could finally get a unified view of at least some aspects of their business. But as these centralized programs grew, they began to exhibit the classic problems of over-centralization. The data warehouse teams became larger and more bureaucratic. Every new data requirement had to go through a centralized approval and development process. What had been designed to enable agility had become a bottleneck, slowing down business initiatives rather than accelerating them. Thus, ironically, disconnected data marts continued to proliferate throughout the landscape, working around the central team.

The predictable response was to swing the pendulum in the opposite direction. Under various names—data mesh[2] being the most prominent—organizations began to decentralize responsibilities. The idea was to push data ownership and management out to the business domains that generated and understood the data best. This approach promised to eliminate bottlenecks and restore agility to data-intensive initiatives.

But going too far in this direction often created its own set of problems, including returning to the splintered data issue that earlier centralization was meant to correct. Even as enterprises sought to integrate their business silos to provide a unified customer experience across business units or operate their supply chains with a holistic understanding of product flow, enterprises that separated data teams into almost completely independent domains made it difficult or impossible to enable such cross functional initiatives with the kind of true data interoperability they would need. Providing general guidance to "federated" teams on how to create data products and contribute them to a shared catalog was not and is not sufficient to solve the problem.

The Modern Data Landscape

Today's data strategy must account for three primary groups in the data ecosystem: data consumers, data producers, and central data teams. Understanding the appropriate role for each group is crucial to finding the right balance between centralization and decentralization of responsibilities.

- *Data consumers* represent the broad community within business domains who need access to data to do their jobs effectively. This includes business analysts, application developers, data scientists, and business users who need self-service capabilities. The goal is to empower these consumers without

[2] For the original article on data mesh, see Zhamak Dehghani, "How to Move Beyond a Monolithic Data Lake to a Distributed Data Mesh," *Martinflowler.com*, May 20, 2019.

overburdening them with data management responsibilities they shouldn't have to deal with.

- *Data producers* are the business domains and systems that generate data as an output of their core operations. These might be sales systems, manufacturing processes, customer service platforms, or financial systems. Each produces data that has value beyond its original operational purpose.
- *Central data teams* provide coordination and shared services that enable the entire ecosystem to function effectively. This includes everything from platform infrastructure to data governance to cross-domain integration services.

The Consumer Perspective: Empowerment Without Burden

The concept of self-service data access is fundamentally sound. Data consumers should be able to explore new data sources, train experimental AI models, build prototype applications, create sandboxes for testing hypotheses, and ask business questions they're authorized to investigate. This kind of self-service capability is essential for maintaining organizational agility and enabling innovation.

However, there's a critical line between empowerment and burden shifting. If data consumers are expected to curate core enterprise data—sales figures, customer records, inventory levels, financial transactions—then the central data organization is no longer simply offering self-service to data consumers; instead, they've abdicated data management responsibilities, relying on the wider community to build (redundant and disconnected) capabilities for themselves. These widely-needed data assets should arrive in a condition that's ready for consumption, not requiring extensive cleanup and preparation by every individual user or application development team.

The appropriate role for consumers in data strategy is to leverage well-managed data assets for their specific business needs while taking responsibility for experimental and exploratory work with new or unproven data sources. They should be empowered to innovate and experiment, but they shouldn't be forced to rebuild foundational data management capabilities that should be provided to them as shared services.

The Producer Perspective: Domain Ownership with Standards

Data producers have deep knowledge of their operational domains and are best positioned to understand the sources, quality characteristics, and business context of the data they generate. A well-designed data strategy should empower these domain experts to take ownership of their data and contribute that data to serve the broader organization.

This means data producers should be provided with the tools, capabilities, and authority to manage their data according to enterprise standards while respecting their domain expertise. They should be able to implement data quality processes, define business rules, establish data lineage, and create documentation that reflects their deep understanding of the data's meaning and limitations.

However, domain ownership doesn't mean complete autonomy. Data producers must operate within enterprise frameworks that ensure their data can integrate effectively with other domains when business requirements demand it. This involves adherence to common standards for metadata management, security, and quality measurement. It also means aligning data values and structures across domains to the extent that consistency is required to join data in a way that makes business sense.

Note that data producer domain teams are almost always data consumer domain teams as well. The distinction refers to the role these teams play in specific circumstances. For example, a sales operations application team produces sales data that is used across the organization, but it also consumes not only its own data, but data from other groups, such as product management.

The Central Team Perspective: Coordination Without Control

The role of central data teams is perhaps the most nuanced and frequently misunderstood aspect of this balance. These teams must provide enough coordination to enable effective data management and cross-domain integration without creating the bottlenecks that led to excessive decentralization as a reaction.

Central teams should focus on capabilities that cannot emerge naturally from decentralized domain teams. This includes establishing enterprise data standards, providing shared infrastructure and services, managing cross-domain data relationships, and ensuring that data can be linked meaningfully across organizational boundaries.

Consider the challenge of customer experience analytics in a financial services company. Individual business units might manage mortgage data, credit card data, wealth management data, and retail banking data as separate domain-owned data products. Each domain team can excel at managing their specific data, but understanding the total customer relationship requires joining data across all these domains.

This cross-domain integration isn't just a technical challenge of putting data on the same platform or providing a unified catalog. It requires ensuring that customer identifiers work consistently across domains, that product hierarchies align appropriately, that transaction dates and amounts can be reconciled, and that the semantic meaning of data elements is consistent when they're combined for analysis.

This level of coordination cannot happen automatically through distributed domain ownership alone. It requires proactive facilitation from central teams who can work across organizational boundaries and ensure that the integration actually works—not just technically, but semantically and mathematically.

And if you're establishing this linkage across domains specifically because it's required by targeted business initiatives, which should be the case, then it's work that would have to be done anyway. In other words, someone will have to do the integration work, so you may as well facilitate it from the central team so the results can be reused across initiatives rather than leaving it to consumer teams to do repeatedly for different use cases.

From Chaos to Coordination: A Telecommunications Transformation

I once worked with a major telecommunications company that was implementing their data strategy using the data mesh philosophy, or at least that's how they referred to it. This was my first encounter with a large-scale implementation of this approach, and it revealed a fundamental disconnect that I've seen many times since.

They had enthusiastically distributed data responsibilities to domain owners throughout the business. The mobile network team owned network performance and tower data. The consumer division owned residential customer data. The enterprise division owned business customer data. The billing team owned revenue and payment data. The customer service team owned support tickets and call center data. Each domain owner was building their own "data products" with impressive technical sophistication.

When I asked how each domain owner decided what to work on, they explained that it was up to each owner to prioritize their data products based on their domain's priorities. "That's the beauty of data mesh," they said. "Each domain has autonomy to serve their stakeholders as they see fit."

But in our discussions about their business vision, they'd revealed something else entirely: The company had one overarching metric driving their entire strategy—revenue per customer through bundled services. They wanted customers to seamlessly combine mobile, internet, streaming services, and home security into integrated packages. They wanted to predict churn across all services and prevent customers from leaving for competitors. Major board-level initiatives were aligned to this convergence vision. And from several interviews, it was clear that they had long struggled with "siloed" data, with no indication that this problem was being addressed.

So, I asked the obvious question: "Who's responsible for integrating customer data across all these domains to enable the convergence initiatives?"

"Well, that's up to the data owners," they said, or something to that effect.

The problem is that this kind of integration does not occur organically when each data owner is focused on their own priorities, and it certainly wasn't happening here.

The mobile network team had usage patterns but not home internet consumption. The billing team had payment history but couldn't connect it to network quality issues that might drive churn. The customer service team knew why people called to complain but couldn't link that to actual service performance or billing data.

Each domain was optimizing locally, but nobody was responsible for creating the integrated customer view that strategic initiatives required.

They saw the disconnect. Here they were, pursuing a company-wide strategy that required understanding total customer relationships across all products and services while their data strategy explicitly avoided cross-domain coordination.

We didn't advise them to abandon domain ownership—the domain teams genuinely understood their data better than any central team could. Instead, we recommended reintroducing strategic coordination. A central team would build a coordinated roadmap aligned to the convergence initiatives, then cascade specific requirements to each domain owner.

The network team would still own performance data, but now they needed to expose it in a way that supported churn prediction models. The consumer division would still own residential customer data, but now they would conform to enterprise customer identifiers that would work across all divisions. The billing team would still own revenue data, but now they knew they needed to structure it to enable bundle analytics across services. And they needed to prioritize their work at least in part based on these shared initiatives. That is, alignment to these shared initiatives would drive which parts of which data domains should be deployed and when, which data quality issues to resolve, and so on.

Each domain owner remained free to pursue domain-specific initiatives. The network team could optimize tower placement, and billing could improve collections processes, for example. But these local optimizations couldn't interfere with the enterprise convergence requirements, and where possible, should leverage the same integrated data structures.

The lesson was this: domain autonomy is valuable, but strategic business initiatives require orchestration. The level of coordination in your data strategy should match the level of integration demanded by your business strategy.

Matching Integration to Business Strategy

The right balance between centralization and decentralization isn't a universal formula. It must be tailored to your specific business strategy and initiatives. Some business initiatives require enterprise-wide data integration, while others can be fully contained within specific departments or business units.

The key is to analyze your portfolio of business initiatives and understand the scope of data integration they require, either within specific business areas, across the enterprise, or both. Initiatives focused on customer experience typically require integration across multiple customer touchpoints. Supply chain optimization might require integration from procurement through manufacturing to distribution. Financial reporting demands integration across all revenue and cost centers.

By mapping your business initiatives to their data integration requirements and observing commonalities across the initiatives, you can determine where centralized coordination is essential and where domain autonomy is sufficient. This business-driven approach prevents both the over-centralization that creates bottlenecks and the over-decentralization that prevents necessary integration.

The Partnership Imperative

Regardless of how responsibilities are distributed between central teams and domain teams, data strategy always requires partnership between IT and business functions. Technical capabilities must be combined with business knowledge at every level of the organization.

This partnership is essential for consumers who need both technical tools and business context to use data effectively. It's essential for producers who must combine domain expertise with technical data management capabilities. And it's essential for central teams who must understand business requirements deeply enough to provide meaningful coordination services.

Moving Forward Iteratively

Finding the right balance between centralization and decentralization is not a one-time decision but an ongoing process of adjustment and refinement. Start with your current business initiatives, implement the minimum level of centralization needed (and no less) to support them effectively, and then adjust based on results and changing requirements. The mandate to integrate data for targeted business initiatives and applications will guide this balance.

The domain teams may report into IT teams segmented by domains, to the business areas directly, or as sub-teams reporting into the central data team. If you're early in developing the discipline to establish cross-domain linkage, then having domain teams as sub-teams within a centralized data team usually works best. As the program matures and you gain confidence that the required discipline is in place, you can safely begin distributing the teams more directly across IT or the business.

The goal is not to achieve perfect balance immediately but to create a framework for making these trade-offs consciously and systematically. By staying focused on business initiatives and maintaining clear guidelines for distributing responsibilities, organizations can avoid the pendulum swings that have plagued data strategy for decades and instead find sustainable approaches that deliver both agility and integration where each is needed most.

One result of effective balance in the organization is effective balance in the architecture. But what should that architecture look like?

That's what Principle #4 is about.

PRINCIPLE #4: BUILD COHERENT ARCHITECTURE, NOT MONOLITHIC DATA STORES OR DISCONNECTED DATA PRODUCTS

- ✓ *Do:* Build semi-independent data architecture components with seamless integration capabilities where needed
- × *Don't:* Build a monolithic platform or disconnected data products

The organizational balance we discussed in Principle #3 has a direct architectural counterpart. Just as organizations have swung between centralized and decentralized teams, data architectures have oscillated between monolithic integration and near-complete fragmentation. Modern enterprises requiring seamless data integration across the company and maximum autonomy for individual teams must synthesize the best of both approaches.

The Architectural Pendulum

As we discussed in Principle #3, in the early days of enterprise data warehousing, the vision was relatively straightforward: one centralized database storing all shared enterprise data. This approach worked well for highly reusable core data such as customer records, product catalogs, financial transactions, and so on. But when business domains couldn't wait for the central team, which was often, they built their own independent data marts. Even when these data marts contained data that would benefit the entire enterprise, the isolation persisted. The result was duplicated data pipelines (extract, transform, and load, or ETL) and overlapping data structures scattered throughout the organization.

Then the pendulum swung to the opposite extreme. The solution to the centralized bottleneck seemed obvious: enable domain teams to build their own "data products" independently. While advocates of this approach did promote interoperability, at least on paper, the reality I've observed across many organizations tells a different story. Domain teams publish data products to a shared catalog while central teams publish metadata standards and general "how to" guides on building data products, but when the data products are built, they can't truly link together in a way that meets business needs for integration.

Yes, they might use the same platform technologies and access the same data lake. But when a cross-functional initiative needs to join sales data with logistics data and customer service data, for example, the integration doesn't work. Different domains use different product identifiers. Temporal boundaries don't align. Critical linkage points are missing. Application teams and analysts end up doing the hard integration work themselves, or they bypass the data products entirely and go back to source systems, just as they did before when the "official" architecture was centralized.

The Performance Reality

Beyond semantic disconnect, there's a fundamental performance problem with excessively distributed architectures. For databases to effectively optimize queries, data that needs to be joined must reside in the same database. This is especially critical for high-volume, complex, frequently-reused data powering multiple applications with significant user activity.

Some believe virtualization can solve this problem—that you can simply drape a virtualization tool over disparate data sources and somehow achieve both semantic integration and adequate performance. I've watched this enticing idea resurface every few years. The appeal is obvious: avoid the work of physical data integration by querying data where it sits.

Here's the reality: virtualization is a great idea for tactical situations like experimenting with new data sources, accessing low-volume reference data temporarily, or creating quick prototypes. But for core analytical workloads involving complex joins across large data volumes, virtualization cannot deliver the performance end users expect. If it could, purpose-built data warehouse technology wouldn't even exist. Yet these technologies are more prevalent than ever because the physics of distributed query execution, network latency, and cross-system optimization haven't changed.

When you need rapid response times for interactive analytics that join millions of transactions with product hierarchies, customer segments, and promotional calendars, you need that data physically co-located in an optimized database. This understanding should strongly influence how we think about architecture.

Two Architectural Paradigms

Two architectural approaches have gained prominence that need to be reconciled: building distributed data products, which enables distributed data ownership and consumption, and implementing layered architectures, which enables cross-domain coherence and performance. The distributed approach, typically under the heading of "data mesh," focuses on domain teams creating data products independently. The layered approach, which processes data through Raw, Harmonized, and Curated layers (or Bronze, Silver, and Gold if you prefer the medallion[3] architecture terminology), has been around for decades and organizes data transformation from across domains through structured progression.

Both have merit. Although each is typically discussed in isolation of each other, you shouldn't choose one over the other. You need both to enable coherent, integrated, enterprise data while effectively distributing development responsibilities across teams. The layered approach works because transforming data from source system structures into integrated enterprise data, and then into application-optimized forms, requires transformation through distinct stages. But going too far toward the layered architecture alone recreates the bottlenecks that plagued earlier data warehousing, causing widespread over-correction toward independent data products, creating fragmentation that prevents true interoperability. And the pendulum swings, back and forth.

The Synthesis: Layered Architecture with Domain Segmentation

The solution is adopting a layered architecture while segmenting it so domain teams can contribute mostly independently. I say "mostly" because achieving true interoperability and good performance requires

---

[3] The medallion architecture terminology was popularized by Databricks, based on an approach to layered architecture that has been used since the early days of data warehousing.

shared architectural elements. The key is letting domain teams own as much of their domain architecture as possible while retaining central elements that enable genuine linkage and good performance.[4]

The Raw Layer: Preserving Domain Reality

The Raw layer captures data as it exists in source systems, preserving all the complexities and domain-specific representations. When different domains represent the same business entity differently, both representations are valid within their contexts. The Raw layer maintains this diversity.

Each domain team ideally owns the ingestion of their source data into the Raw layer. They understand their systems best and can determine optimal extraction patterns, timing, and data structures. The central team provides the platform and standards for how data enters the Raw layer, but domain teams largely control the ingestion processes.

The Harmonized Layer: Creating Integration Points

The Harmonized layer is where architectural coherence emerges—or fails to emerge if not built properly. This is where you establish both semantic linkage across domains and physical co-location of frequently-joined data.

Domain teams build their pipelines through the Harmonized layer, but they don't do so in isolation. The central team establishes and directly helps to implement integration standards that domain implementations must follow:

Conformed master data and facts: The central team develops data models for what's common across domains. These aren't complete models of everything. They're focused primarily on the linkage points between domains such as customer identifiers that work across all domains, product hierarchies that enable consistent categorization, time dimensions that align temporal data, and event structures that capture common attributes.

Domain teams then add their domain-specific content to these common structures. The sales domain adds their unique sales metrics and dimensions. The service domain adds their specific customer interaction data. The logistics domain adds their delivery and routing information. But all of them conform to common customer identifiers,

---

[4] For more guidance on balancing domain and enterprise level data architectures for integration and performance, see Stephen Brobst and Ron Tolido, *An Engineering Approach to Data Mesh*, Teradata Corporation and Capgemini, 2021.

product categorizations, and temporal structures established by the central team, in partnership with the domain teams.

Physical co-location for performance: This is crucial and too often overlooked until performance challenges occur in production, making corrections much more difficult. To address performance effectively, it's important to remember that a database optimizer only works when it has access to all the information it needs to build an optimal query plan. For that reason, high-volume, frequently-joined, commonly-accessed data must be stored in the same database to get the best performance. The central team is usually well positioned to identify which data needs this treatment based on business requirements and actual query and processing patterns.

To achieve this, domain teams may maintain their data in domain-specific databases within the Harmonized layer for domain-specific uses, but they must also contribute to shared integrated databases when their data is frequently joined with other domains.

Standard calculations and metrics: The central team facilitates patterns for how common metrics should be calculated, not mandating single definitions where unwarranted, but ensuring consistency when the same definition for specific elements is needed across domains. For example, "revenue" might be calculated differently for different purposes, but when the finance domain and sales domain both need "recognized revenue," they use the same calculation.

The Curated Layer: Domain-Owned Optimization

The Curated layer returns more control back to domain teams. Here they create data structures optimized for specific applications and use cases. Domain teams own their data products in this layer, exposing them to application teams and end users.

But the Curated layer also includes cross-domain structures that leverage the integration work done in the Harmonized layer. When a business initiative requires true cross-domain analytics—like a supplier scorecard pulling from procurement, logistics, finance, and quality domains—the Curated layer provides these integrated views, built on the semantic linkages and physical co-location established in the Harmonized layer.

And although domain teams may decide what data elements and definitions are appropriate for their specific applications, it's still essential to conform to consistent definitions to avoid confusion across the enterprise. I can't count the number of times I've heard the story of separate groups showing up to a meeting with different numbers for the same metric because each team had their own special twist to calculations, destroying the consistency created in the Harmonized layer.

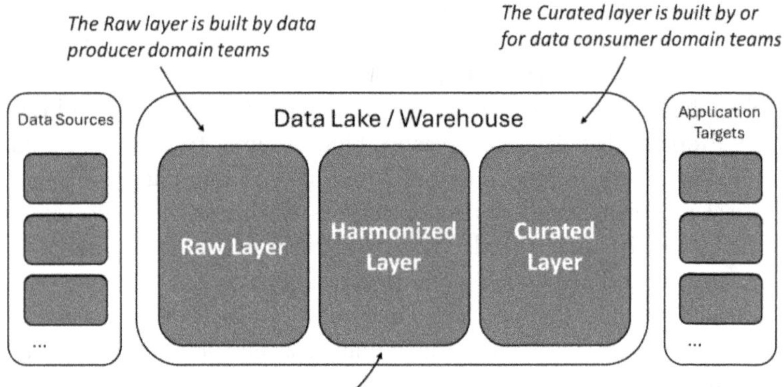

Layered Architecture with Distributed Responsibilities – A layered architecture enables transformation of source data into structures that are useful for target applications while also enabling distribution of responsibilities across teams.

Allocating Architecture Responsibilities

The division of responsibility between central teams and domain teams should maximize domain independence while retaining strong, proactive coordination from the center.

The central team:

- Defines common data models for cross-domain entities (customers, products, locations, time, etc.)
- Establishes integration standards and patterns
- Determines which cross-domain data requires physical co-location for performance
- Provides the technical platform (compute, storage, orchestration, access, etc.)
- Provides data management capabilities (data quality, data lineage, master data management, etc.)
- Facilitates resolution when domains have conflicting approaches
- Facilitates development of cross-domain integrated structures in the Harmonized layer and as needed in the Curated layer
- Monitors and drives overall architecture coherence

The data producer domain teams:

- Ingest their source data into the Raw layer
- Participate in cross-domain integration when required for targeted business initiatives
- Transform their data through the Harmonized layer, conforming to common structures, in partnership with the central team and other domain teams
- Add domain-specific attributes and metrics to common structures
- Ensure data quality (defined as "fit for purpose") within their domain
- Provide ongoing stewardship for their data throughout its lifecycle

The data consumer domain teams:

- Identify and articulate application / use case requirements
- Build Curated layer data products in support of targeted applications
- Aggregate data as needed for targeted applications to meet business needs and enable optimal performance for specific applications
- Reuse data in the Harmonized layer (rather than creating redundant and overlapping pipelines)
- Work with other consumer domain teams and central team to avoid inconsistencies across domains (e.g., defining "revenue" consistently or using a different name for a metric that is calculated differently)

Notice that domain teams build across all three layers, partnering with the central team particularly for the Harmonized layer. While domain teams oversee the complete data journey from Raw to Curated, the actual work often spans multiple teams based on their roles as data producers or consumers.

For example, consider a customer service team building a chat agent application. As data consumers, they create the Curated layer data products tailored specifically for their application, structuring customer and product data to meet their exact needs. Meanwhile, the teams that own customer and product systems—the data producers—are responsible for ingesting this data into the Raw layer and standardizing it in the Harmonized layer.

This division of labor is typical: producer teams handle data acquisition and initial refinement (Raw and Harmonized layers), while consumer teams shape the final data products in the Curated layer. However, producer teams may also build Curated layer products based on consumer requirements, depending on skills and staffing across teams.

To clarify this division further, think of the Curated layer as belonging primarily to the Application workstream according to the seven workstream model we introduced in Principle #1, whereas the Raw and Harmonized layers belong to the Data workstream, with responsibilities distributed accordingly.

When Integration Matters

Remember, you should only integrate where business initiatives require integration. Don't force harmonization across domains that genuinely operate independently. If your European and Asian operations have no shared customers, no integrated supply chains, and no consolidated reporting requirements, don't create artificial integration. Keep them architecturally separate.

But when business initiatives span domains—and many important initiatives do—you need deliberate integration at both semantically and physically. When your business strategy requires unified customer experience across channels, that's when you build harmonized customer data structures. When supply chain optimization needs to see product movement from supplier through manufacturing through distribution, that's when you create the semantic and physical linkages across those domains.

This is yet another reason why it's so crucial to let business initiatives drive the data strategy. If business initiatives tend to exist only within specific business units, then integration of data within those business units is largely sufficient. The scope of integration in the architecture should match the scope of integration expected in the business, no more and no less.

Standards That Enable Rather Than Constrain

The difference between enabling coherence and creating excessively rigid centralization lies in the nature of your standards.

Enabling standards establish interfaces between domains without dictating all internal structures. They're like standardized electrical outlets, allowing diverse devices to connect without mandating internal wiring. For example: every data product exposing customer information must map to the enterprise customer identifier. Domain teams maintain their own identifiers if necessary for domain needs but also provide the mapping to shared identifiers for integration.

Excessively constraining standards force uniformity where diversity has business value. They're like mandating every appliance use the same internal voltage. For example, forcing every domain to use only the enterprise customer identifier internally would lose valuable domain-specific relationships.

The central team should focus on the minimum required standardization that enables required integration. More standardization isn't always the best choice. Appropriate standardization based on business needs is what works.

Building Incrementally

You don't need perfect integration on day one. Build semantic linkages and physical integration as business initiatives require them, following the extensibility guidelines from Principle #2.

Start with a small set of cross-domain integrations that matter for an important business initiative. Perhaps customer experience transformation requires linking sales, service, and digital interaction data. Build the common customer structures, establish the integration patterns, and demonstrate the value through targeted initiatives. Then extend to the next domain when the next initiative requires it.

This incremental approach allows you to learn and adjust. You'll discover which integration patterns work well and which need refinement. You'll understand which physical co-location strategies deliver the best performance gains. Most important, you'll build credibility by enabling the success of important business initiatives that require integrated data and reliable performance rather than asking for faith that someday your comprehensive integration approach will pay off.

The Multiple Data Platform Reality

To be clear, by advocating for deeply integrated data where required, I'm not proposing you build one monolithic database or even one single, unified data platform. A company with independent business units with only lightweight business interaction across units might have separate platforms for each unit, for example. Global organizations might have regional data lakes and data warehouses for performance and regulatory reasons.

You might replicate data across multiple databases when different workloads have different requirements. Customer data might live in a regional database for regional analytics and in a global database for enterprise reporting. You might build an enterprise data lake to enable the Harmonized layer then provide direct access to or replicate segments of that data into region-specific data warehouses in the Curated layer. That's not wasteful redundancy. It's optimization for diverse needs.

The Architecture That Didn't Match Reality: A Global Beverage Company's Lesson

I once worked with a global beverage company that was struggling to make their data architecture work. They had invested heavily in building a "global data lake" through a centralized team, but despite many months of effort, adoption was low and projects were dragging on much longer than expected. The disconnect between IT and the business was palpable, yet no one could quite pinpoint why engagement was so poor.

On paper, the global approach seemed logical. The company was organized into geographic regions, and while each operated somewhat independently, they all dealt with similar data domains including beverages, trucks, retailers, suppliers, and so on. Why not consolidate all this information into one unified global solution?

However, as we conducted interviews across both IT and business teams, a different picture emerged. Each region wasn't just "somewhat independent." They were fundamentally different operations. They ran distinct systems, pursued different business initiatives, and operated under separate management hierarchies with region-specific priorities.

There was no overarching business strategy to align these disparate operations, nor any plans to create one. That isn't a value judgment as it wasn't our objective to evaluate the business strategy, which seemed to be working fine. However, it was an important observation with implications for what the appropriate data architecture should be.

The central data team had based their priorities on what appeared to be common, important data domains, but they hadn't aligned with the actual business initiatives driving each region. They were building a solution for a unified global business that didn't actually exist.

Our recommendation was to decentralize most of the central team and pursue architectures aligned to each region. Instead of one global data architecture, we proposed creating dedicated data teams for each region, maintaining only a lightweight global coordination function.

Each regional team would focus on their region's top business initiatives, building integrated, layered architectures and deploying applications that directly supported those specific priorities, thus providing deep interoperability where needed without the extra effort required for global linkage where it wasn't required by the business.

With this approach, regional business leaders saw the data organization as responsive to their actual needs rather than pursuing abstract global data initiatives. By aligning the data strategy with how the business actually operated, rather than how it might theoretically operate, we changed the direction of a struggling initiative toward one positioned to deliver real results.

Avoiding the Extremes

The goal isn't to constrain domains or return to monolithic centralization. It's ensuring that when the business needs integrated insight—which many funded initiatives will require—your architecture enables it rather than obstructs it.

Your data architecture should reflect a truth about modern business: while operations may be organized in domains, customers and markets experience your business, or at least aspects of your business, as a whole. Your architecture should enable both domain optimization and cross-domain integration where required, not force a choice between them. And it should do so with performance that business processes demand, not promises that virtualization will someday make physical integration unnecessary.

Build layered. Segment by domain. Integrate deliberately where business demands it. That's the architecture that delivers both agility and coherence.

But what happens if you have a massive buildup of legacy solutions that don't conform to this approach? How can you convert an expansive landscape with all the accumulated technical debt into a well-structured architecture and begin serving the business right away?

We'll cover that in Principle #5.

PRINCIPLE #5: FOCUS ON THE FUTURE, NOT ON REWRITING LEGACY

- ✓ *Do:* Support new business initiatives and let legacy fall away
- ✗ *Don't:* Rewrite legacy to "fix everything"

If you're responsible for data strategy in a large organization, there's a good chance you've got a mess on your hands. In recent years, many organizations have let their discipline for deploying enterprise data slip a bit. Some have let it slip a lot. In their rush to deploy valuable applications, project teams developed solutions in relative isolation, collecting the data they need for their own purposes, missing the opportunity to reuse data across applications and analytic use cases. On top of this, well-intentioned "self-service" initiatives have often spun out of control. What were supposed to be prototypes, experiments, or ad-hoc analytics have turned into de facto production solutions, with all the risks and support burdens that go with it.

As a result, data is scattered everywhere—data about customers, sales, products, inventory, and so on—leading to inconsistency, excessive interface costs, inability to adequately protect sensitive data, and challenges linking data together for enterprise initiatives. And, ironically, the desire to build solutions quickly led to a dramatic slow-down of projects, since every project had to spend time and effort to collect, manage, and organize data on their own, regardless of how many other applications needed the same data.

Now you own this mess. What do you do?

The Seductive Trap of the Big Cleanup

If you're like most data leaders, you take the most obvious step: You set about cleaning it up. Maybe you'll take advantage of technology "modernization," such as moving to the cloud and implementing advanced data integration and data management tools, to re-architect, re-design, and re-implement existing data resources to create the well-structured and rationalized enterprise data resources that should have been built in the first place.

Please don't do that.

Think about it: If your goal is to simply clean up the mess, what will you really achieve? After spending literally years reorganizing, de-duplicating, and rewriting complex structures and carefully ensuring that all existing applications continue to function, you'll have exactly the same results you have today. And that's if you execute the transition perfectly.

Meanwhile, the rest of the company will continue to focus on initiatives to address the real business priorities—customer experience, supply chain optimization, manufacturing automation—without the benefit of your help because you're distracted by all that rework. Maybe you'll try to fold in support of new use cases while simultaneously rebuilding all legacy data structures and processes from the ground up. That just introduces more complexity and slows critical business projects while they wait on your data projects. They'll just work around you.

I've watched this pattern play out dozens of times. A new data leader arrives, surveys the landscape, and declares war on technical debt. They launch a multi-year "data modernization" program to rationalize, consolidate, and optimize the entire data landscape. Three years and millions of dollars later, they've successfully migrated less than half of the legacy, the business has built new silos to work around the slow-moving modernization effort, and the data leader is looking for their next role.

The Three-Track Approach

There is a much better way. Instead of turning toward the mess and diving in, you need to parse the work carefully into three distinct tracks, each with its own objectives, timeline, and success criteria:

1. *Business Initiative-Driven New Development:* Building the future-state architecture while enabling new business capability

2. *Technology-Only Migration and Modernization:* Migrating and modernizing technology without changing functionality or deeply re-architecting[5]
3. *Surgical Fixes:* Targeted corrections to specific problems

Let me show you how each track works.

Track 1: Business Initiative-Driven New Development

This is your primary track—the one that should consume 60-80% of your resources and attention. Instead of focusing on the mess, focus your attention where it's needed most: the approved, funded business initiatives that are set to transform the enterprise.

All major business initiatives require data for their success. You should deliver that data, just-in-time and just-enough. You should be able to show how every data element delivered, every data quality issue addressed, and every act of data stewardship contributes directly to specific application or analytic needs within funded business initiatives. And because you'll drive the data deployment, you can organize the data rationally and coherently, ready for reuse across the enterprise.

But what about all that technical debt?

Here's what's going to happen: As you deploy data in support of new business initiatives, outdated data structures will begin to fall away. As a result of the changes within the initiatives, source systems will be replaced, as will applications that rely on the old data structures. Little by little, you can decommission obsolete resources as the new structures support modern versions of retired applications. No need for an archeological dig to decipher the internal complexity of existing systems.[6]

Track 2: Technology-Only Migration and Modernization

But you do still have to deal with some aspects of the legacy directly. Maybe your mainframe licensing is up for renewal at an astronomical cost, your on-premises data center is closing, or your database vendor is deprecating the version you're running. These technology-driven migrations are necessary, and they create their own business value

---

[5] For a further breakdown of migration and modernization strategies, see the "Seven R's": AWS Professional Services, "About the Migration Strategies," *AWS Prescriptive Guidance*, Amazon Web Services.

[6] This is similar to the "Strangler Fig" approach (Martin Fowler, "Strangler Fig Application," martinfowler.com, June 29, 2004), which focuses on the evolution of specific systems, whereas the approach I'm suggesting addresses the wider landscape over time.

through cost avoidance, increased scalability, reliability, and other benefits.

The key to success in technology migration is ruthless scope discipline. Resist every temptation to "improve everything while we're at it." I cannot emphasize this enough: technology migration projects that try to simultaneously deeply re-architect the entire landscape have a failure rate approaching 100%.

When organizations say, "Hey, we're migrating on premises database to cloud-native solutions, let's go ahead and deduplicate all this data, rationalize our data pipelines, clean up the data quality issues, and reorganize the schema to be more logical," they're not planning a migration. They're planning a slow-motion disaster.

Here's what technology-only migration looks like when done correctly:

- *Migrate while modernizing only the technology:* Unless they can be completely decommissioned, move the data structures and pipelines exactly as they are. Yes, even the ugly ones. Re-write only to the extent necessary to move workloads to target technologies while maintaining end user expectations. Your goal is to get off the old platform with minimal risk and disruption.
- *Maintain bug-for-bug compatibility:* If the old system calculated customer lifetime value incorrectly but three applications depend on that incorrect calculation (and may have their own methods to correct the math), your migrated system needs to calculate it incorrectly in exactly the same way. You can fix it later in the business initiative track when you're building new applications. If there is business justification to fix this more urgently, it can be prioritized along with other "surgical fixes" (see below).
- *Automate testing relentlessly:* Since you're not changing functionality, you can verify that every query that worked before produces identical results after migration. This is only possible because you're not trying to "fix" everything.
- *Keep the timeline aggressive:* Technology migrations should be measured in months, not years. If your migration project has a three-year timeline, you're not doing a migration—you're doing a rewrite disguised as a migration.

Track 3: Surgical Fixes

In every legacy environment, there are specific issues that cannot wait for gradual replacement and cannot be addressed through pure migration. These require surgical intervention—targeted, precise, and limited in scope.

Common candidates for surgical fixes include:

- *Security vulnerabilities:* Personally identifiable information scattered across unsecured databases, unencrypted data transfers, or excessive access permissions. You can't wait for these to be naturally replaced. They need immediate attention.
- *Maintenance nightmares:* That one data pipeline that fails three times a week and requires two hours of manual intervention each time. The stored procedure that's 10,000 lines of undocumented, inefficient SQL that only one person (who's about to retire) understands. These targeted problems justify focused rewrites.
- *Regulatory compliance gaps:* New privacy regulations, data residency requirements, or audit findings that must be addressed within specific deadlines. (Although addressing any large-scale regulatory compliance requirements should be proposed as business initiatives.)
- *Performance bottlenecks:* The single query that brings the entire data platform to its knees every Monday, or the data pipeline that can't complete within its processing window anymore.

The key to surgical fixes is specificity. Make a list of the specific issues that need to be addressed and resolve them one by one with targeted and controlled scope. Don't let "fix the security issues" turn into "rebuild the entire data platform with better security."

Each surgical fix should have:

- A specific problem statement
- The business impact or risk of the problem
- Clear success criteria
- Defined boundaries (what's in scope and, more important, what's not)
- A timeline measured in weeks, not months
- An owner who understands the constraint to fix only what's broken

The Integration Challenge

The art of this three-track approach lies not just in keeping the tracks separate but in managing the interplay among them. Here's how they work together:

- *Migrations enable new development:* Once you've migrated off an expensive legacy platform, you've freed up budget that can fund support for new business initiatives and deployed technology that can be leveraged for new, targeted applications.
- *New development can take on some surgical fixes:* If a project is slated to deliver a new application in the near term, supported by required data and pipelines, then that effort can also address (or obviate the need for) any surgical fixes identified in the legacy environment associated with the same data domains and pipelines.
- *Legacy decommissioning accelerates over time:* As new development proceeds, more legacy systems become redundant. This reduces the scope of what needs to be migrated or fixed, creating a virtuous cycle.

Managing Stakeholder Expectations

The biggest challenge with this approach isn't technical, it's political. You need to manage several competing pressures. You may be in a situation where:

- *Finance wants cost reduction:* They see duplicate systems and redundant data as waste to be eliminated. You need to show them that targeted new development will naturally reduce this redundancy faster than a massive cleanup effort.
- *IT wants technical simplification:* Your own team members may be the biggest advocates for "fixing everything." They see the technical debt daily and want it gone. Help them understand that building new, innovative solutions when the business needs them is more satisfying than spending years in legacy cleanup.
- *The business wants everything now:* Business leaders don't care about your technical debt. They want their initiatives supported immediately. This approach lets you say yes to them while still addressing necessary technical issues.

- *Auditors want risk mitigation:* Security and compliance findings need to be addressed. That's what the surgical fixes track is for—demonstrating that you take these issues seriously without letting them derail business value delivery.

(Seemingly) Exceptional Circumstances

There are times when a more aggressive approach to legacy cleanup makes sense:

- *Mergers and acquisitions:* When you're integrating two companies, you often need to rationalize overlapping systems more aggressively than organic evolution would achieve. But this counts as a business initiative, so supporting a merger or acquisition with the appropriate data merging is just a special case of aligning data deployment to new business initiatives. The data deployment roadmap should directly support the merger / acquisition roadmap, application by application.
- *Platform sunsets:* If your vendor is shutting down a platform entirely, you don't have the luxury of gradual migration. But this is still a case of justified migration, and even more reason to address it unencumbered by deep re-architecture work.
- *Massive cost pressures:* If inefficiencies in legacy systems are consuming 80% of your budget and preventing any new development, you might need a more aggressive consolidation strategy. But, again, this is an example of cost-justified migration and surgical fixes, even if major surgery is required in this case.

From Paralysis to Progress: A Consumer Products Transformation

I was working with a well-known consumer products company that was facing a crisis of confidence with their data platform. Years of uncoordinated development had created a monster: tens of petabytes of data storage growing rapidly by the day, with redundant pipelines, duplicated datasets, and no clear understanding of what was where or why it existed.

They had massive clickstream files duplicated dozens of times across different projects. Customer data was scattered across hundreds of datasets with no clear lineage. Data pipelines were so intertwined that nobody dared delete anything for fear of breaking some critical but undocumented dependency.

Meanwhile, executives were demanding progress on major initiatives: using AI to accelerate product development, generating personalized marketing content at scale, and optimizing their complex global supply chain. These weren't optional nice-to-haves. Competitors were already building these capabilities, and they didn't want to be left behind.

The CDO's plan was logical. They had planned an 18-month program to "clean up the mess" by rationalizing the data platform to be ready for AI and advanced analytics.

Eighteen months in the consumer products business might as well be eighteen years. Trends change, competitors move, consumer preferences shift. Waiting eighteen months to start their AI initiatives would be corporate suicide. Meanwhile, the various business areas would proceed anyway, exacerbating the situation by deploying even more redundant data stores and pipelines, this time including a wide array of unstructured data as well.

"What if you didn't wait? Why don't you just start supporting the new initiatives now?" I asked.

The CDO asked how we could possibly proceed given their situation. How could they build new capabilities on this mess? They'd just make it worse.

But that's not what I was suggesting. Instead of choosing between fixing the past or building the future, they could do both, just not in the way they were thinking.

We broke their work into the three tracks I described earlier. For the business initiative track, we'd support the AI initiatives immediately, but build them on new, clean, well-architected data structures. This wasn't adding to the mess. It was building the future state in production, with genuine business value driving every decision.

For the surgical fixes track, we applied the Pareto principle. Rather than trying to clean up the entire landscape, we identified the specific problems causing most of the pain. Those duplicated clickstream files? They were responsible for approximately 30% of all storage growth. Customer data vulnerability? There were tools available to scan the environment to find these specific exposures. We could fix these carefully selected problems in targeted projects.

Here's where it got interesting: The marketing AI initiative needed clickstream data for personalization. Instead of having them create yet another clickstream pipeline, we recommended front-loading the creation of a single, clean, authoritative clickstream ingestion process. The marketing initiative would use it first, but we'd immediately redirect other applications to use it too, allowing us to decommission those dozens of duplicate files.

For the technology migration track, they had several legacy data stores on expensive, outdated platforms. The instinct was to "re-write while migrating" including restructuring the data, cleaning up the quality issues, and rationalizing the schemas. We convinced them to just lift and shift as-is to their new standard platforms. Get off the expensive technology now, then re-architect the data and processes later when business initiatives required it.

The lesson was important: You don't have to fix the past before building the future. By focusing on business value (primarily associated with targeted, funded business initiatives) while maintaining discipline about technical debt, you can transform your data platform through evolution, not revolution.

The Path Forward

Moving away from disjointed data resources that hinder your company's ability to compete, move, and change is an admirable goal. But if there's a way to directly support the most urgent business needs of the company, build trustworthy enterprise data resources, and avoid the muck of indecipherable complexity all at the same time, shouldn't you give that approach a try?

The three-track approach isn't about ignoring problems. It's about solving them in the right sequence with the right level of effort. It's about building the future while managing the present, rather than trying to perfect the past.

Using this overall approach gets you to real value much faster, and it also avoids the archaeological dig necessary to understand every bit of the legacy, which was built perhaps over decades, with the people who built it often no longer around. Modern AI tools can help assess the landscape, but they won't obviate the need for time and effort to carefully evaluate all the legacy structures, processes, and interdependencies.

You'll find that after a couple of years of focused execution across these three tracks, you've delivered more business value, retired more legacy systems, and built more reusable capabilities than you would have achieved with five years of "clean up the mess" efforts.

But how exactly do you execute that business initiative-driven development track? How do you ensure that what you're building truly serves business needs while creating reusable enterprise capabilities? How do you avoid creating new silos while trying to eliminate old ones?

That's what Principle #6 is about.

PRINCIPLE #6: PLAN THE PROGRAM, NOT JUST INDIVIDUAL SOLUTIONS

- ✓ *Do:* Use a disciplined program approach to plan and implement solution projects
- × *Don't:* Build individual solutions without an overarching program plan

Even with clear alignment to named, funded business initiatives, there's still a critical challenge that causes many data strategies to fall short of meeting enterprise-level objectives: the lack of a comprehensive program approach. Organizations often treat each solution project as an isolated effort, hoping that following good data management discipline within each project will ultimately make everything come together. Planning the program holistically makes this result much more likely.

A program approach doesn't mean bureaucratic overhead or analysis paralysis. It means having a disciplined framework that ensures every project contributes to a larger vision while delivering near-term business results. It means knowing not just what you're building today, but how today's work contributes to an extensible foundation to meet tomorrow's needs.

The Four-Phase Framework

The following framework is deceptively simple—four phases that cycle continuously—but don't mistake simplicity for lack of impact. Each phase serves a specific purpose and helps you systematically build a program that makes following all the other principles much easier.

Phase 1: Align - Choosing Your Battles Wisely

Remember the seven workstreams from Principle #1? The Align phase is where you begin populating the first two: Business Initiatives and Applications / Use Cases, while also getting a general understanding of what will be needed in the rest of the workstreams to support the first two. But you're not just making a list. You're making strategic choices that will determine how your data strategy will enable the success of the enterprise. You're also establishing a basis for diving deeper into each of the seven workstreams in subsequent phases.

Selecting Business Initiatives

When first starting or resetting your data strategy, you need to be selective about which business initiatives to support. You can't support them all, and trying to do so will put undue stress on the program just as you're trying to find your footing. Here are some filters to help you pick the right initiatives to support:

- *First filter: Approved and funded initiatives.* As we established in Principle #1, you should support business initiatives that have already vetted their value proposition through normal approval channels. If an initiative can't get funded based on its own merits, it shouldn't suddenly become viable just because you have data strategy funding available. This might seem obvious, but I've seen numerous organizations use their data budgets to fund pet projects that couldn't get approved elsewhere. It never ends well.
- *Second filter: Reusability potential.* Among the approved initiatives, prioritize those that will require highly reusable data and capabilities. A customer experience transformation initiative, for example, will likely need customer and product data that many other initiatives will also require. Supporting this initiative first allows you to build foundational data assets that benefit multiple future efforts.
- *Third filter: Willingness to collaborate.* This is the filter that is rarely discussed, but everyone who's succeeded knows is critical. Which sponsors actually want your help? Which business areas are interested in participating in a coordinated approach versus doing their own thing? Forcing your help on an unwilling

sponsor is usually a waste of time. Yes, if there's a high-visibility, strategic initiative that requires extensive reusable data but doesn't want to cooperate, it may be worth pressing the point with executive leadership. But all things being equal, work with people who want to work with you. Success with willing partners builds credibility that eventually brings the reluctant ones around.

At this program level, you're not just identifying initiatives. You're also sketching out at a high level what supporting data and capabilities you might need. What data domains will these initiatives require? What data quality issues might you encounter? What architectural components will be necessary? You don't need detailed answers yet, just enough understanding to see the big picture.

Selecting Applications Within Initiatives

Once you've identified your target initiatives, you need to identify specific application projects within them. These applications can be any data-intensive application including analytics, AI-enabled applications, agentic solutions, custom applications, or packaged applications. Again, resist the temptation to invent new applications disconnected from the business areas that should sponsor the ideas. Look for projects that are already planned and funded, or likely to be funded, within these initiatives. You will find many projects planned within business initiatives that could use the help of a good, practical data strategy.

If you do propose new applications—perhaps you see an opportunity to leverage modern data, analytics, and AI capabilities that the business hasn't considered—do so in partnership with the sponsoring organization. Have them champion the project and get it approved through their normal channels. Remember, you're there to enable other business areas and their initiatives, not to propose direct value of your work in isolation.

For each application you identify, think through the remaining five workstreams at a high level:

- *Data:* What specific data will this application need?
- *Data Management:* What data quality or integration issues must be addressed?
- *Architecture:* What architectural components are required?

- *Security and Responsibility:* What security, privacy, compliance, ethical, or other policy concerns exist?
- *Operating Model:* What people and process capabilities will need to be in place to manage and leverage the required data, analytics, and AI capabilities?

This isn't detailed planning yet. It's ensuring you understand the full scope of what you're committing to support.

Phase 2: Assess - Facing Reality

The Assess phase is where you gain deeper clarity about your starting point. It's where you map the real landscape as it is today, discovering, perhaps, that customer data lives across multiple systems, revealing opportunities to create unified identifiers and integration points. It's where you identify the specific architectural capabilities in place or planned—whether that's cross-domain integration, real-time data ingestion, or vectorization for AI applications—turning gaps into a clear roadmap for enhancement.

This phase isn't just about documenting problems. It's about understanding what's in place so that you'll know what needs to be done to enable your targeted initiatives successfully. You assess across all seven workstreams, but always with an eye toward action.

Assessing Business Alignment

First, validate that your understanding from the Align phase was correct. Are the initiatives and associated application projects you're planning to support still the right targets? Most important, you assert a commitment to support business initiatives and not slip into "building a data foundation" that is disconnected from specific business initiatives.

This is also where you assess the depth of commitment from your business sponsors. Do they understand what their participation will require (and how they will benefit from your help)? Are they prepared to assign subject matter experts to help define requirements? Will application owners make their source systems available for data extraction?

Assessing Current State Capabilities

For each of the supporting workstreams, you'll need to find the right balance of understanding each of the capabilities at a high level for the overall enterprise while paying special, detailed attention to what in-scope business initiatives and application projects will require for success, refining and extending what you uncovered during the Align phase.

- *For Data:* Don't just catalog what data exists. Assess its availability, completeness, timeliness, and accessibility, especially as required for the near-term applications. Can you access the data you need? How difficult will extraction be? What political or technical barriers exist?
- *For Data Management:* What capabilities already exist for ensuring data quality, managing master data, and tracking lineage, for example? More important, how mature are these capabilities across the enterprise and within the context of solving specific, near-term data challenges associated with target applications?
- *For Architecture:* Map the current technical landscape, but focus especially on what specific technologies you'll need for your targeted applications. If your first applications don't require real-time streaming, don't spend weeks documenting your streaming capabilities (or lack thereof).
- *For Security and Responsibility:* Understand current policies and controls that are in place, paying special attention to any new security, privacy, ethical, or other related requirements your applications might introduce. Will you be handling more sensitive data than before? Will you need to comply with new regulations? What new risks will AI and agentic solutions introduce?
- *For Operating Model:* Assess not just what roles and processes exist, but how well they function. Are there data stewards assigned who can help ensure data quality is adequate to support targeted applications? Are the right roles in place for analytics and AI? Is the operating model for planning and implementation adequately linked to the wider enterprise operating model? (More on that in the next chapter.)

The key to effective assessment is maintaining focus on your targeted initiatives. You're not trying to document everything. You're trying to understand overall strengths and challenges of the wider ecosystem, with a particular focus on what needs to be addressed to make your specific targeted applications successful.

Creating Tangible Actions

The output of your assessment should be a clear, prioritized set of proposed actions tied directly to enabling your targeted initiatives, while setting a vision for enterprise capability.

For each gap you identify, be specific about:

- Which applications are affected
- What the impact would be of not addressing the gap
- What level of remediation is required (perfect isn't always necessary)
- What the rough effort and cost might be

This creates a foundation for realistic planning rather than conceptual ideas that may not be implemented.

Phase 3: Plan - Building the Roadmap

The Plan phase is where everything comes together into a cohesive roadmap. But unlike isolated IT "foundation" planning that might schedule work based on *potential* future use, your roadmap must be synchronized with business solution delivery through support of named business initiatives and associated application projects that *require* the data and associated capabilities you'll implement.

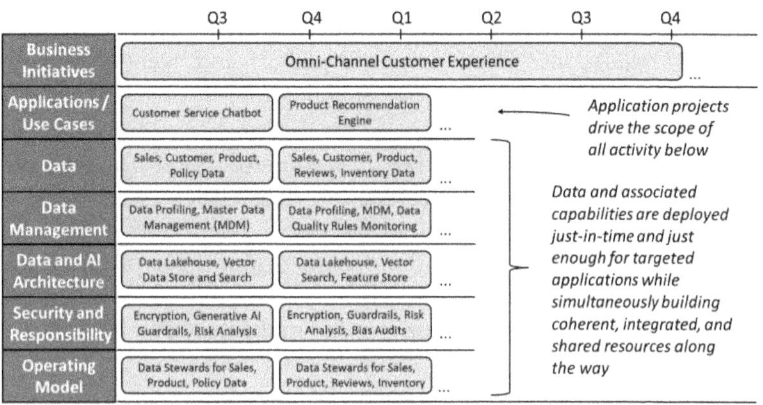

Data Strategy Roadmap – This is a simplified view of a program-level roadmap depicting planned activity across each workstream.

Planning Multiple Iterations

Your roadmap should plan multiple iterations of value delivery, with each iteration accomplishing two things: 1) delivering specific business value through completed applications and 2) building reusable data and associated capabilities that subsequent iterations can leverage and build upon.

Think of it like building a city. You don't build all the infrastructure first and then add buildings. Nor do you build random buildings without infrastructure. Each phase of development delivers usable spaces while reusing and extending infrastructure for future growth.

For example, your first iteration might deliver a customer service chat agent application while also establishing:

- Data integration for sales, customer, product, and policy data
- Basic data quality monitoring for the required data
- A master data management capability for customer and product records
- An initial shared vector data store for unstructured policy data to enable AI-enabled search
- The initial components of your architectural platform
- Basic data stewardship processes for the initial data products

The second iteration might then deliver a product recommendation engine, building on the foundation you've already established, while adding:

- Additional details for sales, customer, and product data
- Integration of product reviews and inventory data
- Expansion of the vector data store for product reviews
- A shared feature store for recommendation model training
- Expanded data stewardship to manage additional data products and data quality issues

Each action noted on the roadmap should be documented in a project charter, epic definition, or whatever mechanism you use to initially scope projects. You may decide to combine multiple interdependent actions across workstreams into individual projects or, especially as the program scales up, you may decide to have separate projects for activity in each workstream. Either way, you'll need to formally define the interdependencies across workstreams without collapsing workstreams into each other.

Detailing Individual Solutions

While maintaining the big picture roadmap, you need to plan at least one solution in complete detail—enough detail to begin implementation immediately. This deeply detailed planning should cover all seven workstreams, but only to the extent needed for that specific solution.

Don't plan data governance capabilities because they're a "best practice." Plan them because without them, your customer segmentation application will produce unreliable results, destroying business confidence. Don't plan architectural components because they're "modern." Plan them because your application needs specific technical capabilities to function. Every element in your plan should trace directly back to enabling the solution. If you can't draw that line clearly in any part of the plan, either the requirements aren't clear enough, or it doesn't belong in your plan.

Sequencing and Dependencies

Pay careful attention to sequencing. While you want to deliver results quickly, some capabilities must be in place before others can be built. You can't implement real-time analytics before you have access to source systems and mechanisms to ingest the data. You can't ensure data quality without clear data ownership and stewardship for associated data domains.

But be careful not to over-serialize your plan. Many activities can and should proceed in parallel. While you're building the technical platform for your first application, you can assess data sources for your second. While you're implementing data quality controls for customer data, you can design the architecture for product data.

Phase 4: Implement - Delivering Value While Building the Foundation

The Implement phase is where you produce tangible results, and sometimes where beautiful plans encounter messy reality. But if you've followed the first three phases diligently, implementation becomes execution rather than exploration.

Building in Slices

Each implementation delivers a complete "slice" of your data strategy—a fully functional solution that delivers business value through one or more targeted applications while contributing data and associated capabilities to your larger foundation. You're implementing elements of all seven workstreams, but only what's needed for that specific solution.

This is fundamentally different from approaches that build applications without any reusable foundational capabilities or build a set of data products before delivering any applications that use the data. You're doing both simultaneously, in careful balance, in separate interdependent workstreams.

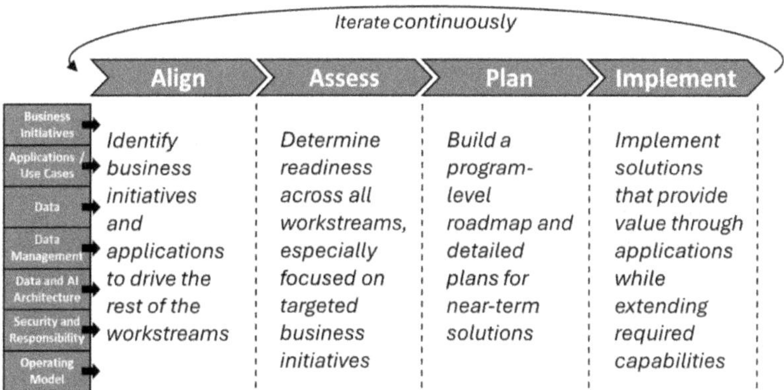

Data Strategy Phases – These phases, while not strictly sequential in all cases, provide a systematic and iterative approach to envisioning the entire program and implementing individual solutions.

The Shortcut to Results

Remember that you don't have to complete all aspects of each phase sequentially. Once you've identified a specific application in the Align phase, you can accelerate through the Assess and Plan phases to the Implementation phase for the most immediate solutions even as program-level assessment and planning continues in parallel.

Maintaining Dual Focus

During implementation, every team member needs to understand and maintain the dual focus we discussed in Principle #2: delivering the specific solution while also building extensible data and capabilities.

The application team focuses on building features that users need today. Data engineers build data structures following the guidelines for extensibility. Technical architects build components that can scale beyond the immediate need. Data governance practitioners establish processes and roles that address near-term issues while accommodating future growth. And so on.

This dual focus requires constant communication and occasional tough decisions. When the application team wants the data team to take a shortcut that would compromise reusability, you need mechanisms to catch and address that. When the data team wants to over-engineer for hypothetical future needs, you need to pull them back to practical delivery.

From Fragmentation to Focus: An Agricultural Transformation

Let me share an example that illustrates how the program approach transforms scattered efforts into coordinated value delivery.

A large agricultural firm I worked with had multiple data-related activities underway, each pursuing its own agenda, each with its own definition of success.

The data governance team had crafted an impressive charter outlining roles, responsibilities, and core policies. Their value proposition centered on "improving data quality," which was a response to a recent internal audit that had correctly identified numerous data quality issues. They'd developed a comprehensive roadmap for establishing governance processes: data quality management, master data management, data lineage, and so on. But they hadn't connected any of their planned work to specific applications or business initiatives. They were planning to improve data quality in general, without knowing which quality issues actually mattered most for business outcomes.

Separately, the IT team was building an "enterprise data lake." They were ingesting data from multiple sources—field sensor information, weather data, crop yield information, soil composition analyses, and so on. Their stated goal was to "make it easier to access data," and they'd identified an initial set of data subjects to implement because they believed these would be widely used. Yet when pressed, they couldn't name a specific application project that was waiting for the planned data. They were building it, hoping users would come.

Meanwhile, the business was launching a major strategic initiative called "precision farming," which involves using detailed field data, predictive analytics, and automated systems to optimize every aspect of crop production. This meant analyzing soil conditions at a granular level to determine exact fertilizer needs for each field section, using weather predictions and crop growth models to optimize irrigation schedules, and leveraging yield forecasts to make better decisions about harvest timing and resource allocation.

Do you see the opportunity here? Do you see how all three of these separate efforts could have been connected? This is a pattern I see all the time. The precision farming initiative desperately needed high-quality, integrated data. The data lake team was building exactly the kind of integrated platform required. The governance team was working on quality controls that would be essential for trusted analytics. Yet these three efforts ran in parallel without any coordination, each duplicating work the others could have provided more effectively.

When we introduced the program approach, we advised them to reorient all their plans to align with the precision farming initiative. This didn't mean throwing away their work. It meant focusing it with laser precision on delivering value through the targeted business initiative.

The data governance team examined the precision farming roadmap and identified which data quality issues would impact the initiative's applications. For the fertilizer optimization application planned for Q2, field boundary data needed to be accurate within one meter, and soil sample locations needed precise GPS coordinates. Instead of trying to improve all data quality everywhere, they focused on these specific, critical quality issues first.

The data lake team reprioritized their data deployment based on the precision farming application roadmap. For the irrigation optimization system launching in Q3, they needed to integrate real-time sensor data with weather forecasts and soil moisture models. Instead of deploying generic "weather data," for example, they implemented parts of the data domain that delivered integrated, analytics-ready datasets specifically for irrigation decisions.

With this new direction, instead of three teams pursuing abstract goals like "better governance" or "easier data access," they became unified contributors to a business initiative that would fundamentally change how the company operated. The precision farming initiative, which had been struggling with data challenges, now had a fully aligned data organization supporting every phase of delivery.

By reprioritizing their detailed work to provide just-in-time and just-enough data and capabilities, these teams didn't just support precision farming. They became integral to its success. And the capabilities they built for precision farming became the extensible foundation for the next wave of agricultural innovation initiatives.

The Continuous Cycle

The four phases aren't a one-time journey. They're a continuous cycle. As you implement solutions, you're already identifying the next wave of initiatives to support. As business strategies evolve, you reassess and replan. Each iteration makes the next one easier and more valuable.

But maintaining this cycle requires more than just good project management. It requires embedding data strategy into the very fabric of how your organization operates. You need mechanisms to ensure that every new business initiative considers data requirements from the start. You need governance that survives leadership changes. You need capabilities that persist beyond individual projects.

So, how do you institutionalize something as complex as data strategy without creating bureaucratic overhead that slows everything down? How do you ensure that the principles and practices we've discussed become "how we do things" rather than just "that data project we did once"?

Principle #7 addresses these questions.

PRINCIPLE #7: EMBED DATA STRATEGY, DON'T ISOLATE IT

- ✓ *Do:* Institutionalize data strategy within the enterprise operating model
- × *Don't:* Institutionalize data strategy in a silo

After all the work you've done, including aligning to business initiatives, building for extensibility, balancing centralization and decentralization, and planning comprehensive programs, there's one final challenge that determines whether your data strategy becomes a permanent capability or just another project that fades away. You need to embed it into the machinery of how your organization operates.

Here's what often happens: Organizations launch their data strategy with great fanfare. They align to business initiatives, build impressive capabilities, and deliver real results. But three years later, when the original sponsors have moved on and new initiatives emerge, everything starts to unravel. The data governance processes that once ensured quality slowly atrophy. The architectural standards that enabled integration get ignored in the rush to deliver. New projects bypass the carefully constructed data resources because it's not perceived to be a faster path, even if it causes overall data coherence across the enterprise to wither.

The problem isn't that the strategy was wrong or the implementation was poor. The problem is that it was never truly institutionalized. It existed as a separate program alongside the business rather than as an integral part of how the business operates.

The Moving Train Problem

Throughout this book, I've emphasized aligning your data strategy to business initiatives that are already funded and planned, and might even be in motion. But there's an inherent challenge in this approach that becomes apparent once you've done it a few times: it's exhausting.

Finding the right initiatives requires detective work. Convincing their sponsors that you can help requires selling skills and political savvy. Adjusting your timelines to match theirs requires flexibility. Coordinating with teams that have already made architectural decisions requires diplomacy. It's like running alongside a moving train and jumping on. It's possible, but hardly sustainable as a long-term approach.

At the beginning of a data strategy program, you have no choice. Those initiatives already have momentum, and if you want to deliver results, you need to catch up and jump aboard. But wouldn't it be better to catch the train at the station like a normal person?

That's what embedding your data strategy into the enterprise operating model accomplishes. Instead of constantly chasing initiatives after they've started, you become part of the process that creates, shapes, and supports them from the beginning.

The Enterprise Operating Model Connection

Every organization has an operating model—a set of processes, governance structures, and decision-making mechanisms that keep the business running. These aren't always formally documented or even consciously designed, but they exist. Strategic planning happens somehow. Projects get funded through some process. Architecture decisions get made by someone. Standards get enforced (or not) through some mechanism.

Your goal is to weave data strategy into these existing processes so thoroughly that it becomes nearly impossible to plan an initiative, fund a project, or make an architectural decision without considering data implications. Not because someone mandates it, but because it's simply how things are done.

This isn't about creating new bureaucracy or additional approval gates. It's about ensuring that the processes that already exist naturally incorporate data considerations at the right moments, with the right people involved.

Let me show you how this works at each level of the organization.

Strategic Planning: Getting in at the Beginning

Strategic planning is where business initiatives are born. Whether it's an annual planning cycle, quarterly business reviews, or ongoing portfolio management, this is where executives decide what the organization will focus on for the coming year or years. If you're not part of this process, you're always playing catch-up.

The ideal situation is having the chief data officer (or equivalent leader) as a direct participant in strategic planning to do essentially three things:

First, become a source of inspiration. The executives planning business initiatives often don't know what's possible with modern data, analytics, and AI capabilities. By participating in strategic planning—even if initially just as an advisor—you can help shape initiatives from the very beginning. You're not just supporting their ideas; you're helping generate better ones. Again, this is much better than proposing these initiatives separately as part of the data program.

Second, observe what emerges. Every initiative that comes out of strategic planning will need data. By being present when these initiatives are conceived, you can immediately begin planning how to support them. You know what's coming before project teams are formed, before budgets are allocated, and before architectures are envisioned.

This early visibility allows you to revise your data strategy roadmap proactively. When the customer experience initiative is approved in October's strategic planning session, you can start laying the groundwork for integrated customer data immediately, rather than scrambling when the project team comes asking for it in February of the following year.

Third, influence the framing. How initiatives are conceived shapes how they're executed. If a customer experience initiative is framed from the beginning as requiring integrated data across all touchpoints, that becomes a fundamental requirement rather than an afterthought. If a supply chain optimization initiative acknowledges at the outset that it needs real-time visibility across suppliers, that proactively drives architectural decisions and the level of centralization and decentralization you'll need to enable this kind of deep, cross-domain data linkage.

Funding Processes: Following the Money

Every organization has some process for allocating resources to projects. This might be an annual budgeting cycle, a quarterly funding review, or an ongoing investment process.

The funding process is another opportunity to ensure that data considerations aren't an afterthought. By the time projects appear in the funding process, they've already scoped their work, estimated their costs, and set their timelines. If data acquisition and preparation are buried inside each project, you've already lost the opportunity for wider reuse and coherence.

Instead, here's what should happen:

When projects are proposed for funding, someone with a data strategy perspective should review them all together, looking for common data needs. It's like the game of Concentration—flip over cards and find the matches. This project needs customer data. That project needs customer data. This other project needs product data. That one also needs product data.

Once you've identified these commonalities, you can propose separate data projects that serve multiple initiatives. This isn't about adding cost. It's about extracting the redundant data work from individual projects and consolidating it into shared efforts. The individual application projects actually become cheaper and less risky because they're not responsible for data acquisition and quality.

Once these connections are established, you need formal dependency tracking between your data projects and the application projects they support. The project management office should recognize and monitor these linkages just as they would any other project interdependency. This creates accountability on both sides—the data team (including central and domain teams) must deliver on time for the applications to succeed, and the application teams must engage with the data team rather than going their own way.

Enterprise Architecture: Building the Shared Foundation

Most large organizations have some form of an enterprise architecture function. Sometimes it's a formal team with detailed frameworks and governance processes. Sometimes it's just a few senior architects who try to maintain some consistency across projects. Either way, this function is crucial for embedding data strategy.

The problem is that many enterprise architecture groups focus almost exclusively on technology standards. Which database should we use? Which encryption mechanism? Which development framework?

While these are important questions, they miss the bigger picture of data as a shared enterprise resource. Your enterprise data architecture shouldn't just standardize technology—it should ensure that data resources are deployed as shared assets.

If your organization has an architecture review board or similar governance mechanism, data architecture should be a standard part of that function. Every project should be asked: What data does this project need? Are you leveraging existing enterprise data resources? If not, why not? What data will this project produce that others might need?

Again, this isn't about creating bureaucratic obstacles. It's about ensuring that the investments you've made in shared data resources get used, and that new projects contribute to the enterprise capability rather than creating more silos, which actually speeds up projects.

Solution Development Lifecycle: Institutionalizing Implementation Practices

Most IT organizations have some standard approach to delivering projects, with varying levels of formality. This methodology typically includes standard phases, activities, deliverables, and roles. This is where you can embed data management practices so they become routine rather than exceptional, particularly within projects that focus on deploying data for use in multiple applications.

Instead of hoping that project teams will remember to profile their data for quality issues, make data profiling a standard task in the methodology. Instead of expecting teams to figure out data governance on their own, define the role of the data steward as a standard project role. Instead of leaving data cataloging as an optional nice-to-have, make it a required deliverable.

Here are some examples of what this might look like in practice:

- During project initiation: Identify required data domains and engage appropriate data stewards. Review existing enterprise data resources for reuse opportunities.
- During requirements gathering: Document information requirements using standard enterprise terminology. Validate requirements against existing logical data models.
- During design: Extend enterprise data models (and domain-specific models linked to common entities in the enterprise model). Design data quality rules based on business requirements.

- During development: Profile source data to identify quality issues that could hinder success of targeted applications. Implement data quality monitoring and remediation processes.
- During testing: Validate that data quality meets specified thresholds. Verify that data lineage is properly captured.
- During deployment: Register new data assets in the enterprise catalog. Establish ongoing data stewardship responsibilities.

The key is making these activities standard parts of the methodology, not special side activities or afterthoughts. Projects can opt out of tasks that are truly not applicable, but the default is inclusion.

Creating the Feedback Loop

Embedding data strategy into the enterprise operating model creates a virtuous cycle. Strategic planning identifies initiatives that need data. The funding process resources both the initiatives and the supporting data and capabilities. Enterprise architecture ensures coherent implementation. The development lifecycle institutionalizes good practices within implementation projects. And the results feed back into the next round of strategic planning.

But this doesn't happen automatically. You need to consciously create and maintain these connections. Here's how:

Start where you have influence. You probably can't revolutionize all these processes at once. Pick the one where you have the best relationships or the most obvious value proposition. Maybe there's a mature architecture review board that's missing data resource oversight. Maybe there's an enterprise or business unit strategic planning event coming up soon, and the chief data officer or designate can participate. Start there and expand.

Document the connections. Make the linkages explicit. If your data quality improvement efforts are supporting three business initiatives, document those dependencies formally and report the results. If your enterprise data sharing approach is enabling faster project delivery, report that acceleration. These documented connections and shared successes prevent your data strategy from being seen as overhead.

Build incrementally. You don't need to embed every data management practice into the methodology on day one, for example. Start with the most critical—perhaps data profiling to identify quality issues early. Once that's routine, add the next practice. Over time, comprehensive data management becomes standard operating procedure.

The Partnership Model

Throughout this embedding process, as with all aspects of data strategy, maintaining the partnership between IT and business is crucial. The strategic planning process needs business leaders who want proactive participation from IT. The funding process needs IT leaders who understand business priorities. The architecture function needs both technical capability and the ability to find common requirements, including data, across projects. The development methodology needs to serve business needs while maintaining technical discipline.

This partnership can't be forced—it must be cultivated. It grows from repeated successful interactions where both sides see value. The business sees that engaging with data strategy early makes their initiatives more successful. IT sees that aligning with business priorities makes their technical work more valued. Over time, the partnership becomes natural, another part of "how we do things."

From Isolated Program to Organizational Capability

Early in my career, when I initiated a data analytics program at a large grocery retailer, I spent a lot of time and effort getting the program off the ground, largely because I was learning at the same time.

After struggling to get funding for a standalone data platform initiative based on the technical value (consolidated data pipelines, rationalized data, and so on), I had finally gained traction by aligning the proposed platform to support the most important business initiatives of the company, including a major supply chain program, a labor planning solution, and a few others. It took some selling, and executive support, to convince the leaders of these initiatives that they should rely on the data we were deploying. It was difficult because the initiatives were already under way, and they understandably didn't want to introduce any additional risk.

But once that hard-won connection was made, I wanted to get ahead of other initiatives as early as possible so I could offer support before project budgets, schedules, and scopes were set and were difficult to change.

All projects need money. So, during the annual planning process where all business areas proposed projects for funding, I reviewed every proposed project, especially the ones likely to be approved, to identify common data needs before teams started building overlapping, redundant data resources. When I found patterns, such as multiple projects needing the same sales, inventory, or product data, I would proactively propose projects of my own to provide the shared data services they could use.

Next, we embedded our data program into the enterprise architecture review process. Any new system or major enhancement required a checkpoint, so along with other questions to determine shared architecture needs, we would ask: Would this project create redundant data? Could it leverage existing data platform capabilities? This wasn't to create an unnecessary hurdle; it was to make sure data reuse was considered before it was too late.

We also integrated with the project management office, registering our data delivery projects as formal dependencies for other initiatives. When an inventory replenishment project depended on our store inventory data project, for example, that dependency was tracked and managed just like any other. This visibility transformed our data platform from an optional resource into critical infrastructure that other projects relied upon.

We also embedded data implementation tasks into the standard software development lifecycle. Rather than treating it as an afterthought, teams now had to consider data architecture during the design phase, before any code was written. Project leaders automatically included data platform integration because it was part of the broader design plan template. A simple checkpoint asking, "How will this project source its data?" prevented countless redundant efforts. There were standard tasks for data modeling, data quality analysis, data lineage, and other activities to ensure that they were included as part of project plans.

By weaving our data analytics program into the machinery of how the organization planned, budgeted, and delivered projects, we moved from being a cost that people questioned to becoming infrastructure that everyone simply expected to work, supporting dozens of production applications over time. That assumption of availability was the ultimate sign we had evolved from an isolated program to true organizational capability.

The Ultimate Test

How do you know when your data strategy is truly embedded? Here are the signs:

Business initiatives assume data support. When executives plan new initiatives, they assume that data capabilities will be available to support them. They don't wonder if they'll need to build their own data infrastructure. They know the enterprise capability exists or will exist in time for their needs.

Projects seek out shared resources. Project teams actively look for existing data resources to leverage rather than building their own. Using enterprise data becomes the easier path, not the harder one.

Practices persist through transitions. When leadership changes, when reorganizations occur, or when priorities shift, the data management practices continue because they're built into how work gets done.

Value is self-evident. You no longer need to justify the data strategy's existence. Its value is demonstrated regularly through the initiatives it enables and the problems it prevents.

CONCLUSION: THE JOURNEY TO SUSTAINABLE DATA STRATEGY

We've covered a lot of ground in these seven principles. We've challenged conventional wisdom about aligning directly to business value, insisting instead on aligning to funded business initiatives that produce value. We've shown how to build for change without trying to anticipate every possible requirement. We've explored the delicate balance between centralization and decentralization. We've demonstrated why coherent architecture matters more than disconnected data products or monolithic platforms. We've explained how to focus on the future while managing legacy pragmatically. We've illustrated how a disciplined program approach multiplies the value of individual solutions. And finally, we've seen how embedding data strategy into the enterprise operating model transforms it from a temporary program into an established capability.

Each principle builds on the others. You can't create a successful program without aligning to business initiatives—this would only accelerate activity in the wrong direction. You can't run a successful program without building for extensibility. If you did, you'd be constantly rebuilding. You can't achieve coherent architecture without balancing centralization and decentralization in the organization effectively because you'd have either bottlenecks or fragmentation. And without embedding the strategy into the operating model, even the best program eventually fades.

The journey from struggling with data strategy to succeeding with it isn't just about having more resources, better technology, or even better data. It's about approaching the challenge with the right principles and the discipline to apply them consistently.

These principles aren't theoretical. They're practical guidance born from decades of real-world experience. They've been tested across industries, through economic cycles, and amid technological revolutions. They work because they align with how organizations actually operate.

The path forward isn't always easy, but it is clear. Start where you are. Pick elements of each principle that address your most pressing challenges. Apply them with discipline. Build on your successes. Learn from your struggles. And gradually, what once seemed impossible—a truly effective enterprise data strategy—becomes not just possible, but inevitable.

The organizations that master these principles don't just manage data better. They compete better, innovate faster, and adapt more quickly to change. In a world where data is increasingly the foundation of competitive advantage, that mastery isn't just valuable—it's essential.

ANSWERS TO COMMON QUESTIONS

Throughout my years as a data strategy consultant, I've encountered certain questions repeatedly—questions that reveal challenges organizations face when trying to plan and implement a data strategy for a large organization. In this appendix, I address the most common ones.

How is data strategy affected by generative AI? Does it change any of the principles?

It doesn't change any of the core principles or the seven workstreams—it just adds additional considerations. In fact, generative AI is causing many companies to have to rethink or develop their data strategy not only because it's necessary for generative AI use cases, but because these initiatives are highlighting data challenges that have been causing issues for some time. Generative AI is providing the motivation to finally address these long-standing challenges.

Instead of thinking of data strategy as a prerequisite to generative AI success, think of it as a co-requisite. You can use the targeted application projects to build what you need for data capabilities just-in-time and just-enough. There's no need to get everything perfect before starting generative AI use cases. Go ahead and start. Just make sure you're considering all the workstreams and all the principles.

Think about a sales advisor application, for example. Data "hygiene" within CRM systems has always been a challenge. But now, think about, for example, a generative AI sales advisor application that uses both structured and unstructured data from the CRM system. Let's say the descriptive information about sales opportunities isn't entered in detail and various customer artifacts are kept in private folders, not shared within the CRM application. The sales advisor application will only be as good as the data you feed into it. Generative AI models are amazing, but they're not psychic. Add this to any issues with the structured data—status of opportunities, opportunity amounts, and so on—and you can see the challenge.

For the seven workstreams, again, they all remain the same, but there are additional considerations for each. The business initiative workstream will include initiatives that are enabled by generative AI. You don't want to think of generative AI strategy as totally separate—it fits right in. Besides, going forward, most major business initiatives will leverage generative AI one way or another.

The same goes for the application workstream. Applications will often include some element of generative AI within them, so these fit naturally within that workstream. The data workstream will now have to include unstructured data along with structured data. Data management must expand to include a broader definition of data quality. Think of the things like standard operating procedure manuals, repair instructions, and so on. These must be correct when using them as input into a generative AI application, either for training or inference, or both.

The architecture workstream will need to include traditional elements like data lakes, data warehouses, and data ingestion, but now it will also need to consider vectorization of unstructured data, LLM gateways, guardrails, etc. The security workstream will need to consider any new issues arising from generative AI, including new regulations and even ethical concerns. The operating model workstream will need to consider the impact on data ownership and data stewardship, which will now include unstructured data. You'll also need to consider the roles of data scientists and how they're organized so that you can effectively distribute AI capability across the organization. There are many more users of AI now, seeking help from experts, and there will be greater demand on these precious resources, so distributing responsibilities carefully is more crucial than ever.

How should we measure success of our data strategy? How should we communicate the measures to executives in a way that will earn ongoing support?

The best way to measure success of the data strategy program is to articulate the contribution the data strategy has made to targeted business initiatives and specific applications. Other measures that are more general are fine, such as overall data quality scores, or surveys that indicate how easy it is for application developers and end users to find the data they need. It's good to measure those things. But they shouldn't be the primary measures of success.

Even if you can't be mathematically precise about the contribution the underlying data platform has made to targeted initiatives and applications, it's still the best measure. Let's say you have a transformational change to your company's approach to marketing—better targeting individuals with relevant offers, for example. Let's say that's a top funded initiative for the company. What you want is for the leader of that initiative to say they couldn't have succeeded without the data you and your team provided (and facilitated, by involving data owners and others from various groups). That kind of story is pure gold for the data strategy and is the kind of thing that really earns support from executives.

The executives care very much about the success of major, funded business initiatives. To the extent you're positioned as a "load-bearing wall" underneath those initiatives, you'll be perceived, correctly, as adding real value to the company. If you instead talk about interesting but less strategic measures, like data quality percentages and maturity scores, you won't get nearly the same support.

If we align to business initiatives as you suggest, won't that cause us to create silos of data? We don't want to have to do a lot of rework later to support new use cases, or worse, create more data silos because we didn't consider all requirements.

No, not if you follow the principles carefully. This is always the concern. Many times when I'm working with a client and they're struggling due to lack of adoption, high costs, low business alignment, and so on (Strategic Mistake #2), I'll suggest they identify one or two major initiatives and go to the program leaders to offer support with the data they need. They sometimes resist this because they feel like this is what got them into trouble in the first place (Strategic Mistake #1) with all the data proliferation they're now trying to correct.

But I'm not suggesting going back to that original mistake of burying data deployment within specific business initiatives. I'm only suggesting that you support the targeted business initiatives and associated application projects to drive the *priorities* of the *shared* data deployment. So, the data deployment should still be a separate workstream, but the application track should have a direct dependency on that workstream.

By keeping the data workstream separate, it can have a dual goal of serving the application projects while also building with extensibility along the way. That should be an explicit objective of that workstream (and the other supporting workstreams). That way, every data deployment project does two things: first, it delivers real value by supporting targeted applications that have been vetted and funded on their own merit, and second, it contributes to a coherent data foundation along the way.

Isn't there some level of work that needs to be done at the enterprise level before building "slices" of data in support of specific business initiatives and applications?

Yes, but you want to keep that kind of work under control. You don't want to go too far with it. For example, it's a great idea to create an enterprise conceptual data model before implementing individual solutions. But it's a bad idea to fully articulate the enterprise logical data model with every attribute and relationship you can think of. You just want enough to get a good outline so that each data delivery can put meat on the bones. It's also a bad idea to build the conceptual data model and then say "OK, now let's populate this." That's where you head toward quicksand (Strategic Mistake #2). Instead, you should populate the model only as needed to meet the needs of targeted applications, again, following the extensibility principles we talked about in Principle #2.

Every workstream should have some level of vision before planning. You should have a reference architecture for the architecture workstream, an outline of the future state operating model you're working toward for that workstream, and so on. But this is just an initial sketch. Things will change along the way, and that's OK. For example, you may have an idea of all the domains that require stewardship, and you might create a matrix to document that. But if you then say "OK, let's go recruit some data stewards and put names in these boxes," you'll run into trouble. First, it will be difficult to get people interested in participating. Why are they doing it? Just because data stewardship is a good idea? Second, the program will start to fizzle out because even the people who do participate will start to fade away as there's little specific contribution they're making to important business initiatives.

It makes sense to align to business initiatives, but what if there are a few, core data products that we know many initiatives will need? Can't we just start by building those data products first?

No, I wouldn't recommend that. While you're right that certain data—customers, products, orders—will be widely needed, deploying that data without connection to specific, named business initiatives creates a trap.

Think about what happens when you decide to "build a customer data product first." Where should you start? Which of the hundreds of customer attributes should you include? You might survey users and get fifty different opinions about what's important. Which data quality issues should you fix? Some applications can tolerate missing email addresses; others can't function without them. What level of data freshness is needed? Real-time? Daily? Weekly? Without specific applications driving these decisions, you're guessing.

I've watched organizations spend many months building comprehensive customer data products based on what they thought would be needed. Then the first real application arrives—a customer churn prediction model for a specific business initiative—and it needs three attributes that weren't included but doesn't need most of what was built. Meanwhile, the data quality issues that really matter for churn prediction weren't addressed because nobody knew that's what would be needed.

Instead, identify the specific initiatives and their applications first. If three different initiatives need customer data, great—build the customer data product to support all three, but let their specific requirements drive the priorities.

Now you know exactly which parts of the customer data product to build first, which attributes matter, which quality issues to fix, and what "good enough" looks like. You're still building a coherent, reusable customer data product—you're just doing it with precision based on real, funded needs rather than theoretical completeness.

The difference may seem subtle, but I've seen it play out dozens of times. Organizations that deploy data without specific initiative alignment end up with impressive-looking assets that sit largely unused. Organizations that build the same data products driven by specific initiatives create assets that are immediately valuable and naturally evolve to become more comprehensive over time.

How do we get the business to participate in our data governance program?

Instead of recruiting a bunch of data owners and data stewards because governing and stewarding data is a good idea with all kinds of value, it's much better to recruit them to solve real problems in support of funded business initiatives. Instead of saying, "Hey, we'd like you to be our data steward for claims data as part of our data governance program," you'd say, "Hey, we'd like you to be a data steward for claims data as part of our data governance program, and here's the first thing we'll need your help with. We've noticed that some of the descriptive elements of the auto claims aren't filled out very well, and we think it will impact the success of the AI-enabled claims adjustment tool we're building. Can you come take a look at the issues and help us come up with some ideas to address this?"

Do you see how that's much more actionable and tangible? Something the data steward would be proud to report they're working on? This is also a much better way to get executive support because the program is then positioned to help the success of the most important company initiatives. That's a much better story than saying "We moved from a maturity score of 2.3 to 4.1!"

What's the best way to get application teams and end users to request the data they need?

It's kind of the wrong question. Instead of instituting a request process for data, like an internal website form or some such, it's best instead to proactively identify data needs yourself—or, that is, someone from the data organization. If you wait for people to request the data they need, it's usually already too late. At least it's too late to deploy the data thoughtfully, in a manner that is extensible and reusable, because now there's an urgency.

I've had many situations where I'm talking to a database designer and they say they'll get a request from an application team for some tables, and they're concerned because they don't have time to optimize for performance, much less consider the wider needs of the enterprise. They think this is a handoff problem, but it's really a planning problem. You need to have a comprehensive roadmap covering all seven workstreams so that there's visibility into what will be needed. This gives the right people on the team the opportunity to plan in advance at various layers.

If you do it right, instead of an application developer asking a database designer for tables, you'll have an application project dependent upon a data delivery project which is fully set up to deliver the data needed by the application, and to do so in a manner that works toward a coherent enterprise vision for data overall.

How should we fund the data platform? Since we'd be deploying data in direct support of specific application projects, we could charge them for the labor and technology required, right?

I'd advise against this approach. While it might seem logical to have application projects pay for the data services they consume, this usually backfires in practice. Parochial funding leads to parochial solutions. When each project pays separately, they'll naturally want "their" data optimized for their specific needs, resisting any requirements (such as applying the guidelines noted in Principle #2) that benefit the broader enterprise.

I've seen this pattern often. A customer analytics project funds customer data integration, so they design it specifically for marketing analytics. Then the finance team needs customer data for credit analysis, but the structure doesn't work for their needs. Since they're paying separately, they build their own customer data pipeline. Now you have two customer data solutions, defeating the entire purpose of a shared platform.

The better approach is enterprise funding for enterprise solutions. The business areas sponsor and fund their initiatives and applications, while the central data organization funds the shared data platform that serves everyone. This ensures the data team can build truly reusable assets without being pulled in different directions by whoever is paying this quarter's bills.

But here's the critical warning: centralized funding without specific initiative alignment is how you end up with Strategic Mistake #2—building a disconnected foundation that nobody uses. I've watched organizations with healthy central funding build impressive platforms that sat unused because they weren't connected to genuine business needs.

The solution is centralized funding *with initiative-driven priorities*. The money comes from the enterprise budget, but the work is still driven by specific, named business initiatives and their applications. This gives you the best of both worlds: the freedom to build coherent, reusable assets and the discipline to ensure those assets actually get used.

ABOUT THE AUTHOR

KEVIN LEWIS has spent more than 25 years in enterprise data and analytics, first building his own successful program, then as a consultant helping hundreds of organizations transform their data capabilities. His contrarian insights have helped companies across every major industry move from data dysfunction to strategic advantage.

www.ingramcontent.com/pod-product-compliance
Lightning Source LLC
Chambersburg PA
CBHW060623200326
41521CB00007B/866
* 9 7 9 8 9 9 3 8 5 8 9 1 3 *